AF541568

Techniques in Veterinary Cytology

NIPA® GENX ELECTRONIC RESOURCES & SOLUTIONS P. LTD.
New Delhi-110 034

Techniques in Veterinary Cytology

Protocol Series-Volume 07

Ratan Kumar Choudhary
Senior Scientist
ICAR-National Research Center on Camel
Bikaner - 334001, Rajasthan
Formerly: Assistant Professor
College of Animal Biotechnology
Guru Angad Dev Veterinary and Animal Sciences University
Ludhiana-141012, Punjab

Yashpal Singh Malik
Joint Director
ICAR-Indian Veterinary Research Institute
Mukteswar-263138, Nainital
Formerly: Dean College of Animal Biotechnology,
Guru Angad Dev Veterinary and Animal Sciences University
Ludhiana-141012, Punjab

Devendra Pathak
Principle Scientist
Department of Veterinary Anatomy
College of Veterinary Sciences
Guru Angad Dev Veterinary and Animal Sciences University
Ludhiana-141012, Punjab

Shanti Choudhary
Research Associate, BSTRI_ Surgery, Loyola University, Chicago, USA
Former Teaching Assistant, College of Animal Biotechnology
Guru Angad Dev Veterinary and Animal Sciences University
Ludhiana 141012, Punjab

NIPA® GENX ELECTRONIC RESOURCES & SOLUTIONS P. LTD.
New Delhi-110 034

NIPA• GENX ELECTRONIC RESOURCES & SOLUTIONS P. LTD.

101,103, Vikas Surya Plaza, CU Block
L.S.C.Market, Pitam Pura, New Delhi-110 034
Ph : +91 11 4386 0225, 9717133558, 9540816132
E-mail: newindiapublishingagency@gmail.com
Website: www.niparesources.com

Print ISBN 978-93-58875-27-0

ebook ISBN: 978-93-58878-11-0

Composed and Designed by NIPA®.

Preface

Cytology refers to the scientific study of cells' structure, function, in situ locale and environment. In this field of veterinary science, cells harvested from various biological fluid samples are examined under the microscope to detect normal and abnormal structures. Therefore, cytology is mainly used to detect disease early, monitor the progression of treatment, detect disease progress, detect cancer, and identify many infectious diseases.

This book, "Techniques in Veterinary Cytology," is designed to provide a comprehensive and practical guide to the fundamental principles and basic procedures of veterinary cytology for students, young investigators, veterinary clinicians and other animal science personnel. Though cytology procedures look simple and easy to perform, the experiment results vary depending on the practice, knowledge of the techniques, and, ultimately, the goal of the study, which is what we are looking for.

This book is structured to take readers through the essential steps of sample collection, preparation, and analysis. Section A, "Collection of Biological Fluids," covers the collection of various sample types, including blood, milk, fine needle aspirates, urine, and saliva, from a range of animal species. Proper collection techniques are critical for obtaining representative samples and ensuring accurate cytological evaluation.

The subsequent sections of this book include processing and analyzing samples. Section B, "Blood," describes the isolation, preparation, and immunostaining of peripheral blood mononuclear cells (PBMCs). Section C, "Milk," details the isolation, staining, and immunostaining of milk somatic cells, providing essential protocols for the study of mammary health. Section D, "Fine Needle Aspirates," guides readers through preparing and staining FNA smears, including Giemsa, Leishman, and immunostaining techniques for identifying proliferating cells. Finally, Section E, "Analysis of Morphology of Cells," introduces Image J, a powerful tool for cell counting and cytomorphological analysis. This section equips the reader with the skills to quantitatively assess cytological samples, enhancing diagnostic accuracy and research capabilities.

Each chapter provides step-by-step protocols, helpful illustrations, and troubleshooting tips to facilitate hands-on learning. This manual is an invaluable resource for students, researchers, and practitioners seeking to master the techniques of veterinary cytology.

Authors

Contents

Section E: Analysis of Morphology of Cells

Introduction

Biological fluids such as blood, milk, fine needle aspirates, urine, and saliva from livestock play a critical role in veterinary medicine, animal husbandry and agricultural research. These fluids serve as non-invasive or minimally invasive sources of valuable diagnostic information, enabling the monitoring of animal health, disease detection, and the assessment of physiological and metabolic states.

Blood, the most widely studied fluid, provides insights into an animal's overall health, immune status, and the presence of infections or metabolic disorders. Milk, particularly in dairy cattle, is crucial for assessing mammary gland health and detecting conditions like mastitis. Detection of milk somatic cells on a routine basis and keeping them low is desirable for milk and milk product quality. Fine needle aspirates (FNA) allow for the examination of specific tissues, such as lymph nodes and cancer masses, offering detailed information on cellular composition and the presence of pathological changes in cell morphology and cellular composition.

Urine and saliva are increasingly recognized for their diagnostic potential. Urine analysis can reveal metabolic and kidney function and provide early indicators of systemic diseases. Saliva, being easy to collect, is a growing focus for non-invasive monitoring, hormonal changes, cancer, diseases and stress.

Together, these biological fluids form a comprehensive toolkit for ensuring the well-being of livestock and pets, improving disease diagnosis and prognosis, and enhancing the productivity and health of animals. Their analysis aids in early disease detection, supports therapeutic decision-making, and contributes to the overall sustainability and efficiency of livestock production systems and pets.

Section A
Collection of Biological Fluids

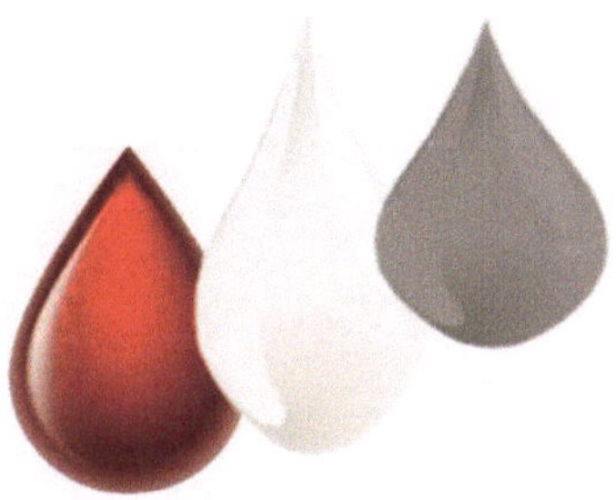

1

Blood Collection from Bovine and Canine

Background

Blood collection from cows and other animals is a routine procedure often performed for diagnostic, research, or health-monitoring purposes. Diagnostic testing, such as hematology, biochemistry, hormone assay, infectious disease testing, and animal pregnancy diagnosis, can be done using blood. Blood samples are required for genetic testing (DNA fingerprinting or parentage determination), pharmacokinetic studies, and immunological studies. Another use of blood collection is monitoring animal health.

It is essential to consider that proper techniques of blood sample collection should be followed. Avoid introducing infection to the animal during blood collection. Remember the animal welfare to minimize distress and discomfort to the animal during blood collection.

Materials Required

- Sterile syringe or vacutainer (usually 16-18 G needle)
- Blood collection tubes (BD Science K2 EDTA Vacutainer(R) PLUS 6 mL) for whole blood, serum tubes for serum collection, or other specific tubes depending on the test (Figure 1).
- Antiseptic solution (e.g., alcohol or iodine)
- Restraint (head gate or halter)
- Gloves

Acknowledments for the contribution to this chapters: Hitesh Rana, Ajayveer Singh and Narinder Kaur Grewal

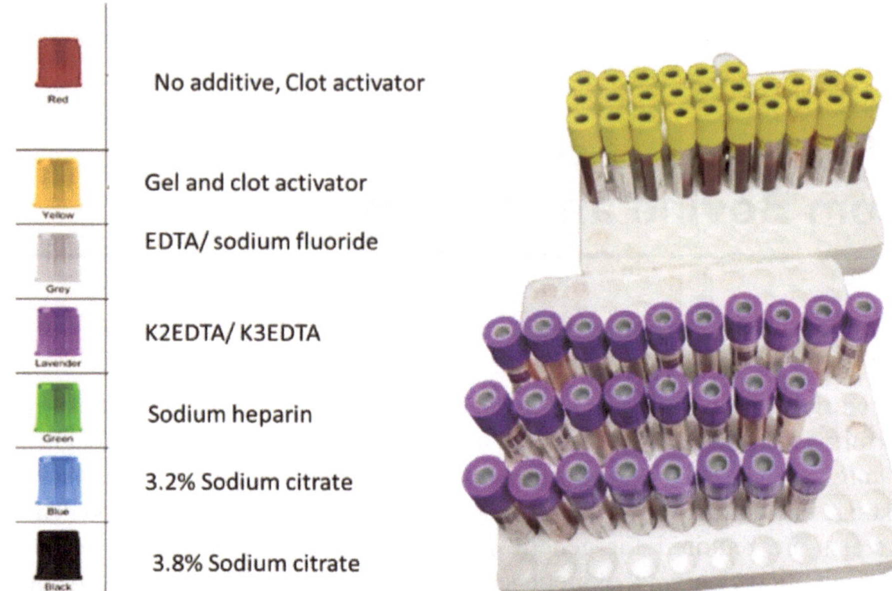

Fig. 1: Different colored caps on blood collection tubes indicate the type of additives inside, which are used for various types of blood tests. Colored caps are fixed and universal. For example, green caps for heparin-containing tubes, blue caps for citrated tubes, red caps for no additives, and so on.

Procedures for Blood Collection in bovine

1. **Restrain the cow**: Proper restraint is essential to keep the cow calm and avoid injury.

   A head gate or halter can be used to secure the animal.

2. **Select the vein**
   - The **jugular vein** is most commonly used because it is large and accessible. It runs along the neck.
   - The **coccygeal (tail) vein** can also collect blood. It is on the underside of the tail, along the midline (Figure 2).

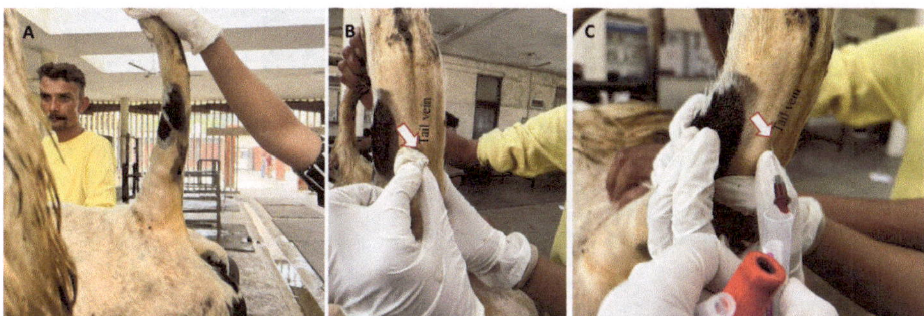

Fig. 2: Blood collection procedure in bovine. Blood is collected from the tail vein (or jugular vein). The animal's tail is raised forward and upward (A), and the tail vein is wiped out of dirt and dung using moist cotton or lint (B). A sterile needle is inserted into the tail vein till blood comes out (C) *(Photo source: Dr. Ajayveer Singh, Ludhiana)*

3. **Prepare the site**
   - Clean the collection site with an antiseptic solution to reduce the risk of infection.
   - If using the jugular vein, apply gentle pressure at the lower part of the neck (near the thoracic inlet) to raise the vein.
4. **Insert the needle**
   - For the **jugular vein**, insert the needle at a 45-degree angle toward the heart.
   - For the **coccygeal vein**, insert the needle perpendicular to the tail, between the vertebrae.
5. **Collect the blood**
   - Draw blood into the syringe or vacutainer until the desired volume is collected. A blood/anticoagulant ratio of 9:1 is optimum for citrate-containing tubes.

Draw about 5-10 mL of blood in nuclease-free tubes containing anticoagulant citrate or K2. EDTA. Gently gently invert the tube 2 to 3 times for proper mixing of anticoagulant and blood. Do not shake vigorously, as this may cause hemolysis. DO NOT INVERT RED TUBES; these tubes separate serum (Figure 3).

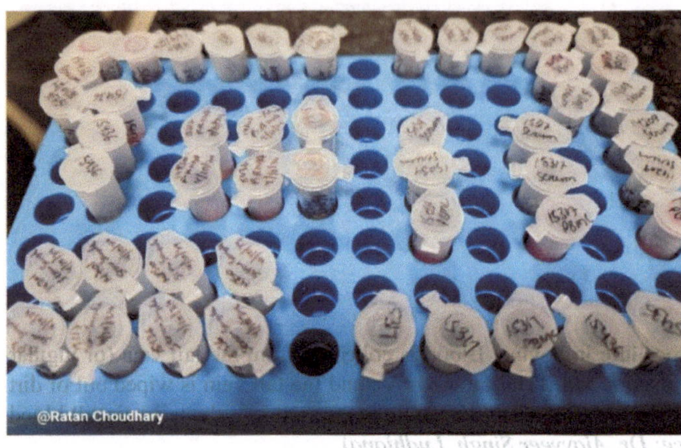

Fig. 3: Storage of serum samples in deep freezer for downstream applications later. Labeling: Label the collection vial with a permanent marker with the following information.

- *Place of collection -*
- *Animal no.*
- *Species – Buffalo*
- *Sample – Blood*
- *Date of the collection*

Samples can be kept up to 5 days at 4 °C or sooner. Transport the sample to the lab in a cool pack. Keep sample tubes straight to ensure that cells should be covered by plasma.

Procedures for Blood Collection in Canine

In dogs, the cephalic or jugular vein draws not more than 1% of blood volume. The dog should be lying down on a table and properly restrained. At the collection site, the dog's forelimb is shaved at the site of collection (away from the paw; Figure 4) over the cephalic vein using a clipper, and 70% ethanol is rubbed. Using a 21 G or wider needle, about 5 mL of blood is withdrawn. After collection, blood flow should be stopped by applying pressure with a sterile gauze for about 30 sec or till the oozing of blood stops.

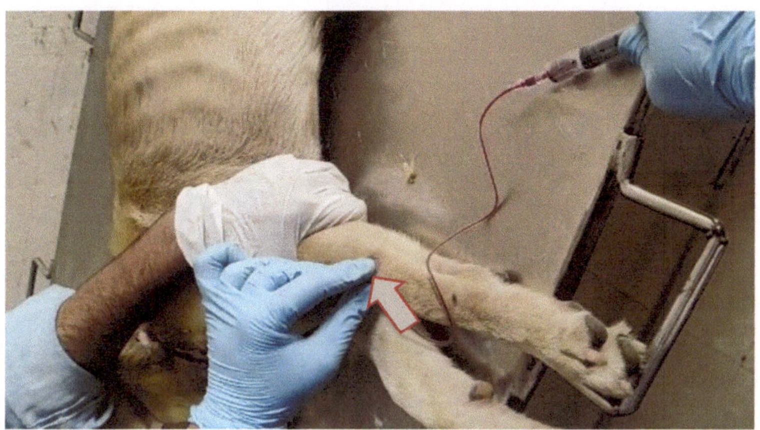

Fig. 4: Blood collection procedure in canine. Blood is collected from the cephalic vein of the forelimb. Not more than 1% of total blood should be collected at a time. *(Photo source: Dr. Ajayveer Singh, Ludhiana)*

Technical Hints

- Avoid shocks (eg; rapid shaking) to the sample to prevent lysis of RBC.
- The total blood volume of a dog is 80 ml/kg body weight or 8% of the body weight.

This means not more than 7.5% of the blood should be collected within 24 hours from the dog. If 10% of blood is withdrawn, then the dog must be given rest for 10 days before the next time blood collection.

Further Readings

UBC, 2020. Cattle blood sampling. Accessed on Oct 17, 2024. https://dc.sites.olt.ubc.ca/files/2021/01/SOP-Cow-001-Cattle-Blood-Sampling.pdf

UMD, 2019. Recommended standard method of blood collection: Canine. Accessed on Oct 17, 2024. https://www.umaryland.edu/media/umb/oaa/oac/oawa/guidelines/blood-collection/Blood-Collection-Guidelines_Canine_07.2019.pdf

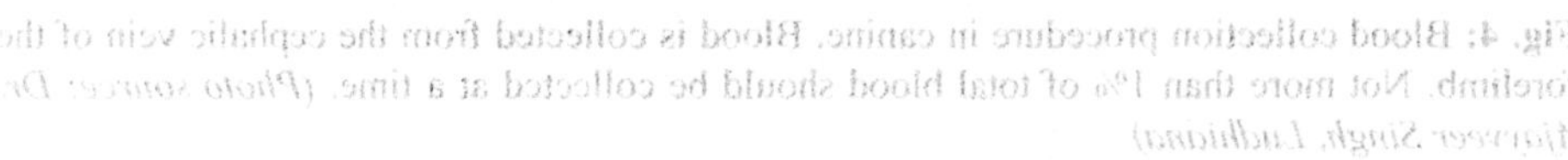

2

Milk Collection from Cow

Background

Milk is known as a whole food for human consumption. It is the first food of the newly born human mammals. Milk contains almost all the nutrients required for the patients, the elderly, pregnant mother and new born baby. There is no doubt that milk plays a significant role in human development. We can prepare many physical forms of milk like cheese, yogurt, ghee, and milk powder. The calcium content is higher in buffalo than in cows. Buffaloes are the second largest source of milk and produce 15 % milk. Higher biological value proteins, fat, soluble vitamins, fatty acids, amino acids, water-soluble vitamins, and calcium are all present in milk. The main antioxidant components of milk and milk products are sulfur, phosphate, zinc, and amino acids like cysteine (necessary to produce collagen).

Milk somatic cells are normal cells found in milk in small amounts (epithelial cells), some of which are leukocytes (white blood cells). When mastitis is infected or injured, leukocyte production rises. Milk with a low SC level produces superior milk products that last longer. Better milk management, better cleanliness, and correct nutrition all contribute to a decrease in milk SC, which indicates poor milk production hygiene and a cause of mastitis. Milk somatic cells are made up of both immunological and milk-producing cells. In terms of defense, somatic cells are the second line. During the routine process of milking, these cells are secreted in milk. Somatic cell count is a good tool for evaluating farm cleanliness in bulk milk and is utilized worldwide as an indication of udder health in individual milk. The most prevalent condition affecting dairy buffalo production and their health and well-being is mastitis. The somatic cell count consists of immune cells (75%), such as neutrophils, macrophages, lymphocytes, erythrocytes, and epithelial cells (25%). Investigators have made different attempts to exploit milk somatic cells as a potential non-invasive sample to diagnose early pregnancy udder health in bovines.

Materials and Reagents Required

- Sterile graduated centrifuge tubes (15 or 50 ml)
- Potassium permanganate solution (5%)
- Nitrile gloves
- Tissue paper roll
- 70% ethanol wipes
- Permanent marker for tube labeling
- A polystyrene container with ice to maintain temperature at 4°C for processing and /or transport to the processing laboratory.

Procedures

Cleaning Steps Prior to Milking

1. First, tie the hind legs of the cow.
2. Freshwater wet the udder region along with the cow's teats to loosen dirt particles.
3. Scrub debris or dung particles stick to the teats or udder of the cow.
4. After ensuring total debris-free cleanliness of the teats and udder.
5. Wash the udder along with the teats with 5% $KMnO4$ solution.
6. After 4 to 5 minutes, wipe the teats and udder with clean tissue paper.

Milk Sample Collection

1. Put on sterile nitrile gloves and apply 70% ethanol to disinfect your hands.
2. Label tubes appropriately prior to milk collection using a permanent marker.
   - Write cow ID on the Q1 tube
   - Quarter # (Q1, Q2, Q3 Q4)
   - Date of collection and time
3. Open the lid of the milk collection tubes and hold the cap facing down (to avoid any dirt falling inside the cap).
4. Hold the lid facing down (ensure the inside of the cap should not touch your gloves), and in the same hand, hold the sample tube at 45 degrees from the teat.
5. Keep the sample tubes near the teat and gently draw milk directly into the tubes to avoid froth formation.

6. Place lids immediately, keep tubes inside the ice container and transport them immediately to the lab for down-streaming applications.
7. In the case of composite milk sample collection, that is, the milk from all four quarters in the same tube, start by sampling from the teats closest to you and then proceeding to the teats on the far side of the udder, collect 1 to 2 ml of milk from each quarter.
8. Store milk samples immediately on ice or inside the refrigerator. Samples to be culture at a later date (more than 48 hours) should be frozen immediately.

Further Readings

https://www.mcgill.ca/research/files/research/dc-615_milk_sampling.pdf

McGill, 2023. Milk sampling. Accessed on Oct 17, 2024. https://www.mcgill.ca/research/files/research/dc-615_milk_sampling.pdf

NMC, 2004. Procedures for Collecting Milk Samples. Accessed on Oct 17, 2024.

3

Collection of Fine Needle Aspirates (FNA) from Canine Mammary Tumors

Background

Fine needle aspiration (FNA) is a minimally invasive diagnostic procedure used to obtain cells from a mass or lesion for examination under a microscope. It's a valuable tool in diagnosing tumors in canines (Figure 5) and other species of animals, including humans.

Advantages of FNA

- Minimal invasiveness to the animals. Thus, no sedation of animals is required.
- Rapid results - Results are obtained faster as cell processing is not required.
- Cost-effective
- Risk-free
- FNA can be performed under image guidance, such as ultrasound, CT, and MRI, for precise identification of the location of cells inside the tissue.
- May be able to differentiate between benign and malignant tumors
- Can assess tumor grade
- FNA helps monitor treatment response. FNA is helpful as it can help pre-plan the treatment of the dogs. If the tumor is malignant, then the treatment can be given immediately, and if it is benign, then the treatment can be postponed.

Limitations of FNA

- Insufficient research materials – often, the volume of FNA samples collected is inadequate for a definite diagnosis. Repeated biopsies may require a large volume of samples for the analysis.

- Sampling error – missing the cancerous part of tissue while performing FNA often results in misdiagnosis.
- FNA may not provide sufficient information about the tumor, such as invasiveness, malignancy, prognosis, and other details.

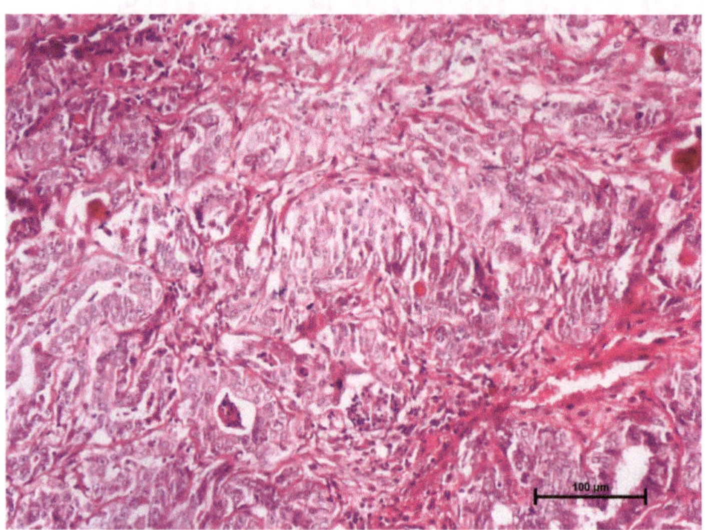

Fig. 5: Histological section of canine mammary tumor stained with hematoxylin and Eosin showing pleomorphic cells with loss cytostructure of mammary alveolar and ductal systems.

Materials and Reagents Required

Fine needle aspiration (FNA) is a minimally invasive procedure used to collect cells or fluid from tissues, typically for diagnostic purposes. The materials required for FNA collection include:

- Needles – 22-25 gauge (G)
- Syring – 3- 5 ml capacity
- Syringe holder (like Cameco syringe holder)
- Gauge and bandages
- Antiseptic solution

Procedures

1. Firstly, the animal is sedated, and a 22-gauge needle attached to a 3-5ml syringe is used to collect aspirates (Figure 6).
2. Needles are inserted into the tumor mass where the presence of the tumor is detected, and then a back-and-forth movement is applied to the

syringe in different directions so that aspirates are collected properly.

3. The aspirates are directly pumped out of the syringe on a clean glass slide to make a smear.
4. To obtain sufficient samples for two smears, generally, 15 to 20 needle passes are required.

   If excessive blood is drawn, limit the needle passes to approximately five to avoid diluting the sample. A stationary needle typically yields only blood.
5. Collected FNA biopsy can be utilized to make smears on the glass slides and identify cancer cells by veterinary pathologists.
6. FNA biopsy can also be preserved appropriately for nucleic acid isolation for genomic and transcriptomic studies to identify tumor subtypes. It can also be used for protein isolation and further downstream applications.

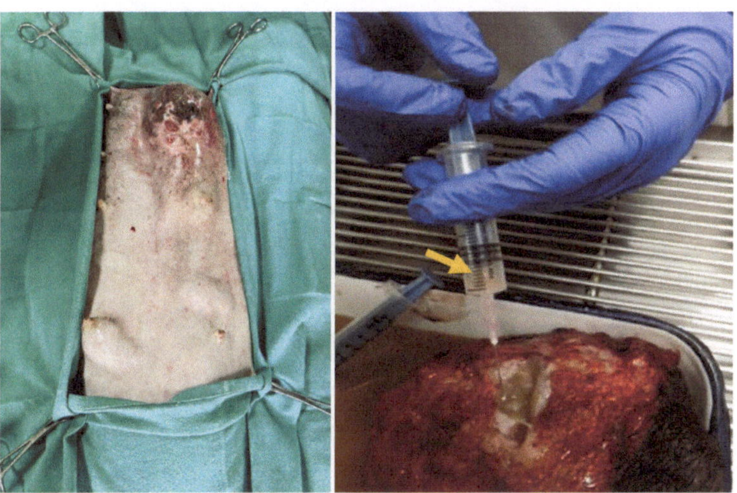

Fig. 6: A clinical case of canine mammary tumor and collection of fine needle aspirates (FNA) from surgically excised tumorous gland. ***Left side*:** Preparation of surgical sectioning of canine mammary tumor is done under general anesthesia. *Right side:* Tissue fluids collected in a syringe obtained from FNA are shown by an arrow (*Photo source: Luhach, Priyanka, 2022).*

Further Readings

Komal Basra and Ratan K Choudhary (2022). Canine Mammary Tumor Fine Needle Aspirate Cytology and Identification of Tumor Types. Acta Scientific Veterinary Sciences 4(6): 102-107.

Ljung, B-M. Techniques of Fine-Needle Aspiration, Smear Preparation, and Principles of Interpretation. Accessed on Aug 21, 2024 https://basicmedicalkey.com/techniques-of-fine-needle-aspiration-smear-preparation- and-principles-of-interpretation/

Luhach, P. (2022). Investigation of fine needle aspirates from canine mammary tumors – a translational model for diagnosing human breast cancer (MSc thesis). Guru Angad Dev Veterinary and Animal Sciences University, Ludhiana.

Sontas, B. H., Yüzbaşıoğlu Öztürk, G., Toydemir, T. F., Arun, S. S., & Ekici, H. (2012). Fine-needle aspiration biopsy of canine mammary gland tumours: a comparison between cytology and histopathology. Reproduction in domestic animals = Zuchthygiene, 47(1), 125–130.

4

Collection of Urine from Animals Including Rodents

Background

Urine collection is a fundamental procedure in veterinary medicine, providing valuable insights into an animal's health status. Urine analysis can detect urinary tract infections, kidney function, and metabolic disorders. Several methods can be employed for urine collection in dogs, each with its advantages and limitations. Always collect urine in clean, sterile and leakproof containers.

Procedures of Urine Collection In Animals

1. Free catch or voided urine: Collecting urine as the animal voids naturally. This procedure is non-invasive and easy to perform. However, the collection of voided urine may not be free from contaminants. Contamination risk from external genitalia and dilution or concentration variation in urine composition is common.
2. Catheterization: Inserting a catheter into the urethra to obtain a sample, often used in larger animals or when free catch is not possible. It provides a sterile sample that is suitable for bacterial culture and cytology. However, catheterization is an invasive procedure with the potential for trauma and discomfort to the animals.
3. Cystocentesis: Inserting a needle directly into the bladder through the abdominal wall to collect a sterile sample. Like catheterization, cystocentesis provides a sterile sample, ideal for bacterial culture and cytology.

The impact of the method of urine collection may affect results. For example, the method used to collect urine from dogs significantly affects the types of bacteria found in the sample. Cystocentesis vs. midstream voiding method of urine collection in dogs resulted in different bacterial compositions. The observed richness of microbiota was higher in urine collected via midstream voiding than in cystocentesis (Coffey et al., 2023).

Urine Collection from Rodents

Mice and rats are used as model animals for the evaluation of acute, chronic toxicity, and metabolic toxicity drugs. Mouse or rat urine can be collected using clear plastic wrap. A plastic wrap can be placed on top of white paper outside the animal cage. The voided urine, each time, can be collected into microcentrifuge tubes.

Alternatively, urine collection can be done over a Petri dish with mild intervention. Keep mice over the clean Petri dish or plastic disposable weight boat and encourage them to micturate. A gentle pressure on the trans-abdominal region over the bladder can induce micturition and approximately 30-100 μl of urine can be collected.

Another method of urine collection from rodents is using a small beaker. Attache a 5 mL of polystyrene beaker to the perineal wall region of a rat using double-sided sticky tape. Hold the rat by the tail using one hand while stimulating the back of the animal using another hand.

The majority of the time, 100-800 μl of rat urine can be collected in the beaker within a few seconds (Kurien et al., 2004).

Further Readings

Coffey, E. L., Gomez, A. M., Ericsson, A. C., Burton, E. N., Granick, J. L., Lulich, J. P., & Furrow, E. (2023). The impact of urine collection method on canine urinary microbiota detection: a cross-sectional study. BMC Microbiology, 23(1), 101.

Kurien, B. T., Everds, N. E., & Scofield, R. H. (2004). Experimental animal urine collection: a review. Laboratory animals, 38(4), 333–361.

5

Collection of Saliva from Dogs

Background

As a non-invasive and relatively stress-free method, saliva holds great potential as a diagnostic sample. It has already been utilized in diagnostic tests for rabies and other diseases in dogs. Saliva secretion is the secretion of salivary glands that maintains homeostasis in the oral cavity, protects against pathogens, and helps digest food. It contains inorganic substances, cells (blood and epithelial cells) and microorganisms. Saliva also contains immunoglobulins, enzymes, hormones, and cytokines derived from serum and bronchial secretions. The saliva of dog is being used to detect rabies virus (Kasempimolporn et al., 2000), parvo virus (Kocaturk et al., 2021), cancer and measure stress levels. Obtaining saliva rapidly, painlessly, and uncomplicatedly can be an easy tool for disease screening.

For the identification of serval prognostic biomarkers for various physiological and pathological conductions, saliva proteomics can be very helpful. In humans, saliva contains more than 3000 proteins (Kawas et al., 2012). Several diseases like oral carcinoma, AIDS, hepatitis B, cystic fibrosis, and other diseases like diabetes mellitus. Differential expression of dog saliva proteins compared to human saliva revealed 2532 proteins. Proteins related to apoptosis and adhesion were predominant in the oral proteome of dogs, in addition to proteins related to regeneration and healing processes such as FGF and EGF (Sanguansermsri, et al., 2018).

Collection Method

Swabs: Collecting saliva using cotton swabs or specialized saliva collection devices. Sterile sample collection tube with appropriate preservatives, depending upon the biomolecules that need to be preserved, either nucleic acid-DNA/RNA or proteins.

Cotton with flavors like citric acid or sucrose may be used to increase the volume of saliva from dogs. It was seen that citric acid generated a greater volume of saliva and greater palatability in dogs, though the pH of saliva would turn acidic.

Procedures of saliva collection (swab method)

- Prepare the dog in a calm and relaxed position.
- Upon wearing protective gloves, insert the swab into the animal's mouth
- Gently rub the swab for 30 sec against the upper lip to expose the buccal mucosa.
- Collect and keep the sample in the sterile tube, ensuring proper closure.
- Label the samples with the following information

  Place of collection - Dog identification no. - Species/breed -

  Sample type – Salivary swab

  Date of the collection –
- Store the swab at the refrigerator or freezer, depending on the type of analysis.

Procedures of saliva collection (Cotton roll method)

The saliva sampling procedures using the cotton roll method in dogs (Damián et al., 2018)

- Picking a cotton roll with a mouse-tooth hemostat and positioning it near the dog's snout for 3 seconds.
- Allowing the dog to smell it. If the dog licked the cotton roll, it was held closer to the mouth to encourage further licking.
- If not, keep the cotton roll inside the dog's mouth to contact the oral and vestibular cavities.
- All the dogs bite the cotton roll at least once and place it back in the mouth until it absorbs saliva.
- After one minute, the cotton roll was transferred to a nylon mesh inside a 15 mL falcon tube.
- Repeat the procedure twice. Keep tubes on ice and centrifuge at 3500 rpm for 15 min at 4°C to collect saliva.
- Finally, the collected saliva was stored at -20 ° C.

Further Readings

Al Kawas, S., Rahim, Z. H., & Ferguson, D. B. (2012). Potential uses of human salivary protein and peptide analysis in the diagnosis of disease. Archives of Oral Biology, 57(1), 1–9.

Damián, J. P., Bengoa, L., Pessina, P., Martínez, S., & Fumagalli, F. (2018). Serial collection method of dog saliva: Effects of different chemical stimulants on behavior, volume and saliva composition. Open Veterinary Journal, 8(3), 229–235.

Kasempimolporn, S., Saengseesom, W., Lumlertdacha, B., & Sitprija, V. (2000). Detection of rabies virus antigen in dog saliva using a latex agglutination test. Journal of Clinical Microbiology, 38(8), 3098–3099.

Kocatürk, M., Tecles, F., Yalçın, E., Cihan, H., Tural, M., Levent, P., Cansev, M., Cerón, J. J., & Yilmaz, Z. (2021). Changes in choline and cholinesterase in saliva of dogs with parvovirus infection. Research in Veterinary Science, 134, 147–149.

Sanguansermsri, P., Jenkinson, H. F., Thanasak, J., Chairatvit, K., Roytrakul, S., Kittisenachai, S., Puengsurin, D., & Surarit, R. (2018). Comparative proteomic study of dog and human saliva. PloS One, 13(12), e0208317.

Section B: Blood

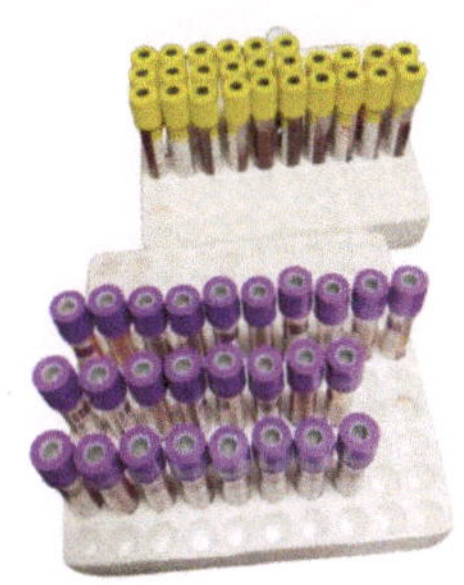

6

Isolation of Peripheral Blood Mononuclear Cells

Background

Peripheral blood mononuclear cells (PBMCs) are a crucial immune system component. Isolating them from whole blood allows for an in-depth study of immune responses, disease mechanisms, and the development of novel therapies. Lymphocytes, monocytes, and other different white blood cells (WBCs) possessing circular nuclei make up the cells of PBMCs. These cells can be directly obtained from whole blood using different separating methods. PBMCs can be utilized for multiple purposes, from drug assessment to expression analysis to parentage determination.

- Toxicity assessment of novel compounds: Interaction of drugs with PBMCs can help assess the safety of compounds and set appropriate doses for safe consumption.
- Comparative analysis: In vitro assessment of diseased vs. healthy PBMCs can provide insights into affected biochemical pathways. The effects of foreign chemical compounds can also be assayed to drive future studies.
- Personalized medication: Studying PBMCs can help develop a genetic immune profile, leading to more specificity and effectiveness in personalized medicines.
- Occupational disorders: Diseases developed due to the overexposure of certain toxic compounds can also be determined by testing PBMCs in peripheral blood.
- Gene expression analysis: Different biological changes within the organisms can be assessed using differential expression analysis.
- Parentage determination: DNA isolated from the PBMCs can be exploited for parentage determination and whole genome sequencing to predict evolutionary relationships among the species.

Acknowledments for the contribution to this chapter: Paramjeet Sharma and Hitesh Rana

The primary step in investigating the behavior of lymphocytes *in vitro* is to isolate them. Lymphocytes are localized in blood, lymphoid organs, and peritoneal exudates. Peripheral blood can serve as an easy reservoir for isolating lymphocytes. Pure lymphocytes can be obtained by utilizing different separating procedures. Overall, the amount of PBMCs is less than that of other components in whole blood, and it is also challenging to study PBMCs in association with other components of blood. So, PBMC isolation is a must-step before conducting any experiment.

PBMC can be isolated in two different methods

1. **Density gradient method**
2. **RBC lysis method**

Materials and reagents required for PBMC isolation

1. Blood samples (Heparin, EDTA, or citrate may be used as anticoagulants for the blood sample).
2. Sterile graduated centrifuge tubes (15 or 50 ml),
3. Pipettes,
4. Pasteur pipettes,
5. Sterile isotonic balanced salt solution (like HBSS) or any other isotonic solution free from Ca^{2+} and Mg^{2+} ions.
6. HiSep™ LSM 1077 (Density gradient medium, Himedia)
7. RBC lysis buffer of choice
8. Centrifuge with swinging rotor.

Procedures of PBMC isolation using a Density gradient method

1. Dilute blood with sterile isotonic phosphate buffered saline in 1:1 (Figure 7A) Note:
   - Avoid using high-binding tubes, as it may lead to a low yield of cells due to binding with the tube.
   - For good results, process samples within two hours of collection.
   - Prepare dilution 1:1 or 1:2 of whole blood in sterile isotonic phosphate buffered saline (may increase the dilution to 1:4 for absolute cell number).
2. Gently layer the diluted blood sample over a density gradient medium (HiSep™ LSM 1077) in 3:1 (3 parts diluted blood and 1 part density gradient medium) (Figure 7B). Note:

- Warm density gradient medium to 20°C before use. Layer diluted blood over medium with very gentle release from the Pasteur pipette by touching the tip of the pipette to the tube.
- Both diluted blood and density gradient medium should be used at room temperature.

3. Place layered tubes into the swinging rotor holders of the centrifuge with care.
4. Centrifuge at 400 x g for 30 min with brake off.
5. After centrifugation, buffy coat deposits over a density gradient medium (Figure 7C).
6. Collect the buffy coat cells in a fresh, nuclease—and pyrogen-free 15 ml falcon tube and wash twice with fresh DPBS at 600 x g for 10 min (at 20°C).
7. PBMC cell pellets will be deposited at the bottom of the falcon tube (Figure 7D).
8. The supernatant will be carefully discarded without disturbing the cell pellet.
9. Wash the mononuclear cell isolate with DPBS or HBSS followed by centrifugation (600 x g for 10 min) at room temperature (20°C).
10. Resuspended cells in a suitable medium can be used for downstream purposes.

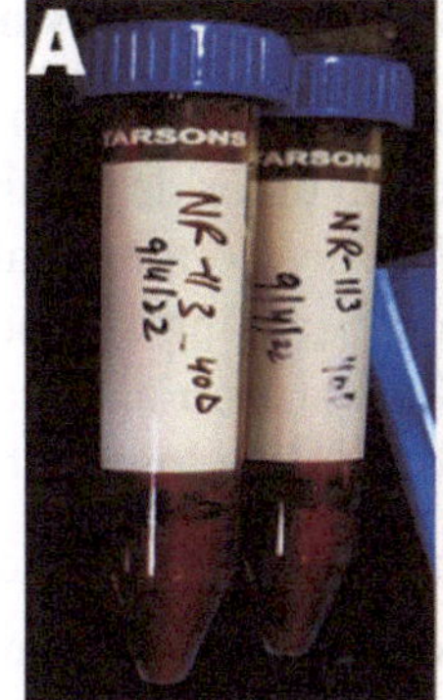

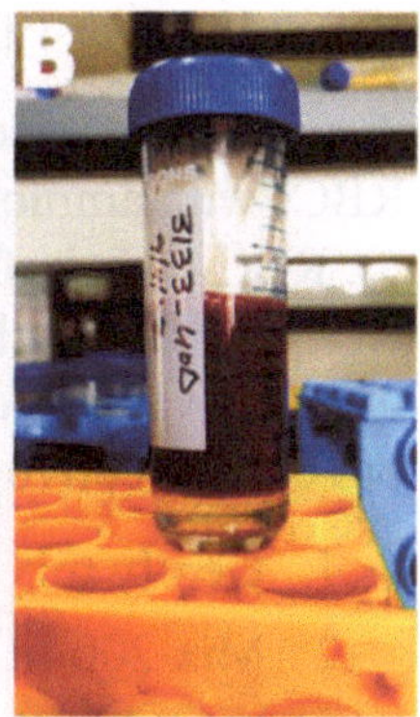

Fig. 7: Layout of PBMC isolation: (A) Diluted Blood Sample, (B) Blood layered on density gradient medium, (C) Buffy Coat, and (D) PBMC Pellet.

Note

- Count the number of viable cells using Trypan blue exclusion staining.

Troubleshoots

- Use fresh and clot-free blood for separation; for the finest results, use blood immediately. Decreased viability and low cell number results due to delayed processing. Cell yield from heparin-treated blood will decrease discernibly after 2 hours, and that of EDTA-treated blood will decrease after 6 hours.

- Removing excess plasma with the buffy coat may contaminate cells with protein.
- Analyze the viability of cells using Trypan blue exclusion staining. In case of low viability, change the washing media to PBS.

Procedures of PBMC isolation using the RBC lysis method

A commonly used method to obtain leukocytes from whole blood is to lyse RBCs with ammonium chloride potassium solutions (RBC lysis buffer; Figure 8). RBC lysis buffers are commercially available and can be prepared to lyse RBCs in whole blood, leaving leukocytes intact. Harvested leukocytes can be used for various downstream applications. The RBC lysis method is manual but cost-effective due to the low cost of the RBC lysis buffer.

1. Gently mix the defibrinated blood sample at room temperature.
2. Mix 200 μl of whole blood sample with 2 ml 1X RBC lysis buffer.
3. Incubate for 5 minutes at room temperature.
4. Centrifuge the RBC lysed sample at 400 x g for 10 minutes.

   Note: Proceed centrifugation with brake off to reduce chances of cell loss
5. Discard lysis buffer by gentle decantation without any jerk.

   Note: *Soak the last traces of lysis buffer on sterile tissue paper to avoid potential harm to the leukocytes suffering from buffer contamination.*
6. Repeat if RBC contamination is observed in leukocyte pellets.
7. Harvested leukocytes are ready for downstream applications.

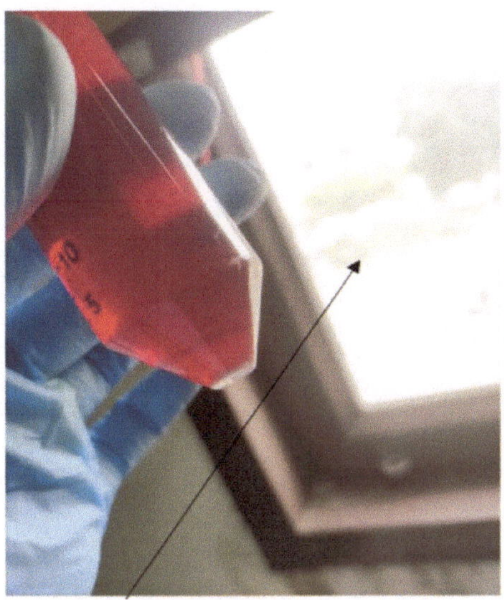

PBMC pellet from whole blood (2.5 mL) after RBC lysis

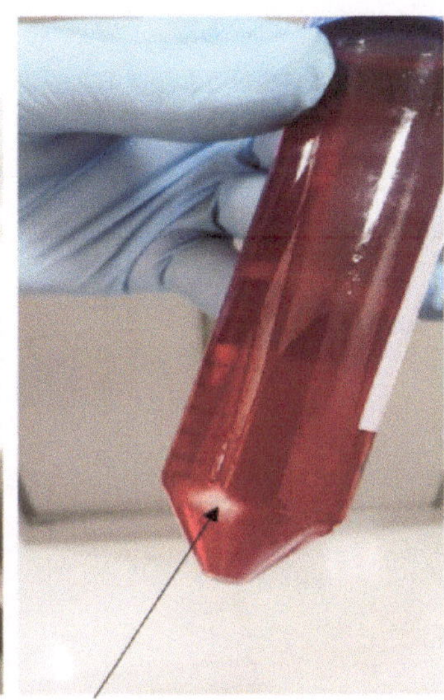

PBMC pellet after removing plasma and RBC lysis

Fig. 8: Isolation of PBMCs using the RBC lysis method.

Further Readings

Böyum A. (1968). Isolation of mononuclear cells and granulocytes from human blood. Isolation of monuclear cells by one centrifugation and of granulocytes by combining centrifugation and sedimentation at 1 g. Scand J Clin Lab Invest Suppl. 9777-89.

Recommended Standard Method for Isolating Mononuclear Cells. https://www.sigmaaldrich.com/IN/en/technical-documents/protocol/clinical-testing-and-diagnostics-manufacturing/hematology/recommended-standard-method (Accessed on Aug 20, 2024)

Sharma P, Choudhary RK, Ratta NS, Singh ST. Investigation of conceptus stimulated gene expression in buffalo peripheral blood mononuclear cells as potential diagnostic markers of early pregnancy. Journal of Dairy Research. 2023;90(2):142-145.

7

Preparation of PBMC Smear

Background

Peripheral Blood Mononuclear Cell (PBMC) smears are a valuable diagnostic tool in veterinary medicine, offering insights into various animal diseases. By isolating and examining PBMCs— comprising lymphocytes and monocytes, veterinarians can detect hematologic abnormalities, immune responses, and the presence of infectious agents or neoplastic cells.

This method is particularly useful in diagnosing viral, bacterial, and parasitic infections, as well as immune-mediated disorders and certain cancers. PBMC smears provide a quick and cost- effective way to assess animal health, aiding disease diagnosis, monitoring, and treatment planning. Following is the utility of PBMC smear in animal disease diagnosis;

1. Identification of abnormalities in WBC morphology
2. Identification of abnormalities in the number of cells like leukosis leukopenia.
3. Certain viral infections, such as canine distemper or feline leukemia virus (FeLV), can cause characteristic changes in PBMCs that are detectable on a smear.
4. Detection of blood parasites like Babesia and Ehrlichia. These parasites are detected using a blood smear; isolation of PBMC from whole blood is not required.
5. Certain types of cancers, like leukemia or lymphoma, can be identified through changes in PBMCs. Abnormal proliferation of lymphocytes or the presence of blast cells can suggest hematologic malignancies.

As we examine the complexities of PBMC smear preparation, we will experience the variation and complexity that make this procedure both an art and a science. With care and attention to detail, we can unravel the mysteries of cellular activities and immune responses present in the blood, making the PBMC smear an important part of the pursuit of a better understanding of animal and human health.

Materials and Reagents Required

1. PBMC suspension in PBS
2. Clean glass slides (Figure 9)
3. Kimwipes
4. Cell suspension of PBMC
5. Fixatives (either methanol or ethanol or 10% neutral buffered formalin or 4% paraformaldehyde)
6. Coverslips (optional)
7. Microscope

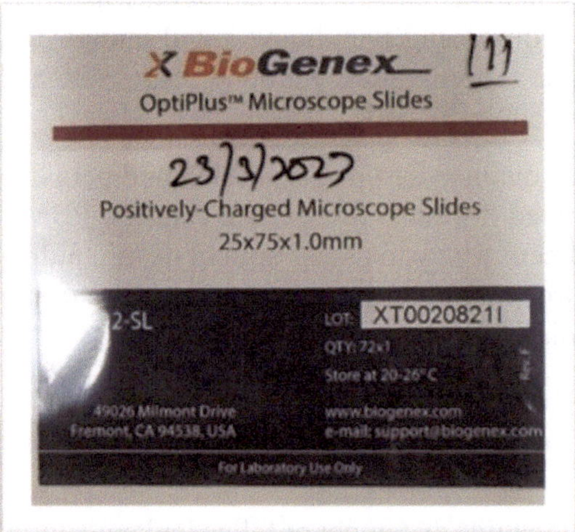

Fig. 9: Positively-charged glass slides can be used for both smear staining and taking tissue sections for immunohistochemistry.

Procedures

There are various methods to prepare PBMC smears. Each individual may prefer their methods for success. When beginning, try different methods to find which works perfectly for you. PBMC smears are made using freshly isolated PBMCs from whole blood or prefixed cells preserved in ethanol.

1. Using Kimwipes, gently wipe two glass slides to remove dust or dirt fragments. Keep the slide on an even surface.
2. Resuspend cell suspension thoroughly. Place 100 μl of cell suspension on one end of one glass slide. Hold the slide's lower end (frosted edge) with the thumb of your off- hand.

3. Using your primary hand, place the edge of the second slide at an angle of 35°- 45°on the first slide, in front of the cell suspension drop. Using gentle pressure, carefully pull the second slide slightly back into the cell suspension drop and allow the cell suspension to spread to the edge of the slide.
4. Push the second slide forward carefully to prepare the cell smear rapidly. It is crucial to use gentle and equal pressure throughout the complete procedure, and do not lift the second slide before it reaches the end of the first slide.
5. After drying, fix the cells by immersing the slide in a fixative solution such as methanol or acetone for 2-10 minutes.
6. Air dry slides again
7. Apply the chosen staining reagent to the fixed smear. Common stains include:
   a. Wright-Giemsa Stain: For general morphology.
   b. H&E (Hematoxylin and Eosin): For tissue sections and specific cell types. c. Antibody Stains: For specific markers (e.g., immunofluorescence or immunohistochemistry). For immunostaining of PBMC- see the next chapter.

Further Reading

Barcelo, H., Faul, J., Crimmins, E., & Thyagarajan, B. (2018). A Practical Cryopreservation and Staining Protocol for Immunophenotyping in Population Studies. Current Protocols in Cytometry, 84(1), e35.

8

Immunostaining of PBMC for Macrophage Identification

Background

Immunocytochemistry is the immunostaining of a specific cell or its components in a cell smear. It is a technique to visualize a specific protein or an antigen in cells based on an antigen- antibody interaction (Figure 10). The site of antibody-antigen binding can be visualized by color development (enzyme-chromogen system) or fluorescence. The macrophages can be visualized in the smear of peripheral blood mononuclear cells by immunocytochemistry using antibodies against macrophages. We need to use a light or fluorescent microscope to detect color or fluorescence, respectively.

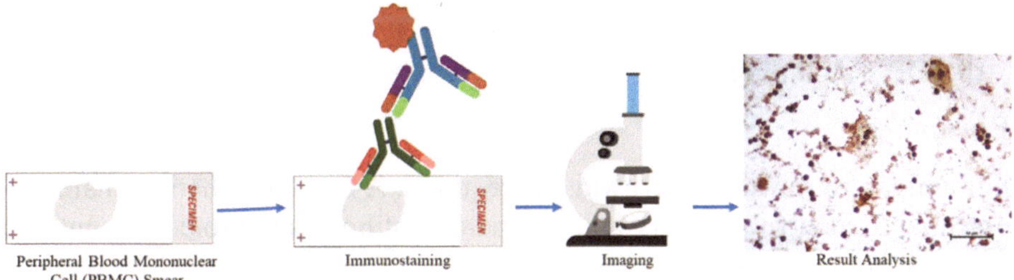

Fig. 10: Overview of Immunocytochemistry

Materials and Reagents Required

1. **Fixative:** The peripheral blood mononuclear cell (PBMC) smear can be fixed using cold acetone, ethanol, or aldehyde-based fixative (neutral buffered formalin, formaldehyde, or paraformaldehyde).
2. **Washing buffer (phosphate buffer saline)**
3. **H_2O_2:** Hydrogen peroxide will be required for endogenous peroxide blocking in the cells in the case of the horseradish peroxidase (HRP) enzyme-based color detection system. Usually, 3% H_2O_2 is used for 20 minutes.

4. **Antigen retrieval buffer:** if the smear is fixed in an aldehyde-based fixative (Formalin, formaldehyde, or paraformaldehyde), then an antigen retrieval will be required. For antigen retrieval, citrate buffer or EDTA buffer is used along with heat (heat-induced epitope retrieval).
5. **Normal horse serum or any other serum:** 2.5% horse serum or any other serum can block non-specific epitopes.
6. **Primary antibody:** primary antibody against the macrophage, either monoclonal or polyclonal antibodies, can be used to identify the macrophage. On the market, both ready-to-use antibodies and concentrated antibodies are available. If a concentrated antibody is to be used, then a perfect dilution of the antibody has to be decided by using multiple dilution trials.
7. **Anti-secondary antibody:** The secondary antibody should be raised against the animal where the primary antibody has been grown.
8. **DAB chromogen**: 3,3'-Diaminobenzidine (DAB) is the most used chromogen for immunocytochemical staining. In the presence of the peroxidase enzyme, DAB produces a brown precipitate that is insoluble in alcohol and xylene.
9. **Haematoxylin:** Haematoxylin is used to stain the nuclei of the cells to get a complete picture of the cells.
10. DPX mounting media.

Equipment

For antigen retrieval, a microwave is required. A moist incubation chamber is essential to incubate primary and secondary antibodies to protect the antibody solution from drying.

Procedures

The staining protocol can be divided into the following steps:

1. **Antigen retrieval**: If the smear has been fixed using an aldehyde fixative, antigen retrieval is necessary. The slides with smears are placed in the citrate buffer for heat- induced antigen retrieval and heated at 95–98 °C for 10 minutes.
2. **Peroxidase block**: Endogenous peroxidase enzyme should be blocked by incubating the sections in 3% H2O2 solution for 20 minutes.
3. PBS washing (3x 3 min).
4. **Protein block:** Non-specific reactions should be blocked by incubating the section with bovine serum albumin or horse serum for 30 minutes.

5. **Primary antibody**: incubation with primary antibody (pre-diluted or concentrated antibodies) can be undertaken inside a moist chamber for 1 hour at room temperature or overnight at 4°C. If the antibody is concentrated, the optimum titer should be worked. An optimum titer will provide a good reaction with minimum background staining.
6. PBS washing (3x 3 min).
7. **Secondary antibody:** incubation with secondary antibodies (pre-diluted or concentrated antibodies) can be undertaken inside a moist chamber for 30 minutes at room temperature.
8. Slides are washed thrice in 0.1 M PBS for three minutes each.
9. Application of DAB chromogen for 2 minutes.
10. Wash for 5 minutes in running tap water.
11. Counterstaining in a hematoxylin solution for 5 minutes.
12. Wash in running tap water for 10 minutes.
13. Sections are dehydrated in ascending grades of alcohol.
14. Smears are mounted permanently in the DPX solution.

Results: The antigen-antibody interaction is visualized by the brown precipitate of DAB using a light microscope (Figure 11). Usually, two types of reactions are seen based on the protein's location: cytoplasmic or nuclear.

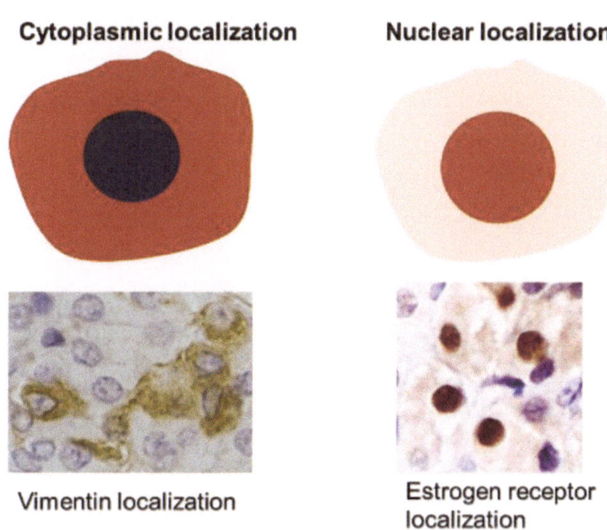

Fig. 11: Immunohistochemical location of vimentin and estrogen receptor in tissue sections. Similarly, smears can be used in place of tissue sections, and the localization of proteins inside the cells can be visualized using immunocytochemistry.

Further Readings

De Matos, L. L., Trufelli, D. C., De Matos, M. G. L., & da Silva Pinhal, M. A. (2010). Immunohistochemistry as an important tool in biomarkers detection and clinical practice. Biomarker insights, 5, BMI-S2185.

Skoog, L., & Tani, E. (2011). Immunocytochemistry: an indispensable technique in routine cytology. Cytopathology, 22(4), 215-229.

Section C: Milk

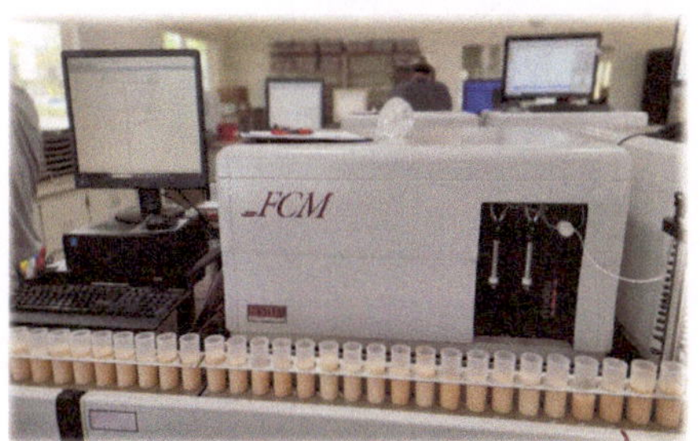

9

Isolation of Milk Somatic Cells

Background

Milk somatic cells are a diverse population of cells found in milk, including epithelial cells, leukocytes (white blood cells), and macrophages. Isolating these cells from milk provides valuable insights into mammary gland health, mastitis detection, and milk quality assessment.

Several methods can be employed for milk somatic cell isolation. Among the others, centrifugation is the most common method of cell separation. Filtration, immunomagnetic separation and flow cytometry are the other methods of isolating specific cell populations.

In bovines, isolating milk somatic cells is useful in milk quality assessment, mastitis detection, and mammary gland biology research. An increased milk somatic cell count level, particularly neutrophil, indicates mammary gland inflammation. Healthy and mastitis glands can be diagnosed by analyzing the differential counting of milk somatic cells.

Procedures for Isolation of Milk Somatic Cells using Centrifugation

1. Collect milk aseptically in nuclease and pyrogen-free containers (Figure 12.
2. Keep it on ice, transfer it to the lab, and place the sample at 4°C.
3. Centrifuge milk sample at 3000 Xg for 15 min at 6°C.
4. Milk fat will get deposited at the top, and the cell pellet will settle at the bottom.
5. Discard fat and skimmed milk carefully, leaving the cell pellet intact at the bottom.
6. Dissolve the cell pellet in 2 ml of isotonic PBS (1X) and transfer it to a new 15 ml falcon tube.
7. Now, entrifuge tubes at 3000 Xg are used for 10 min at 6°C.
8. Cell pellets will be deposited at the bottom of the tube (Figure 13).

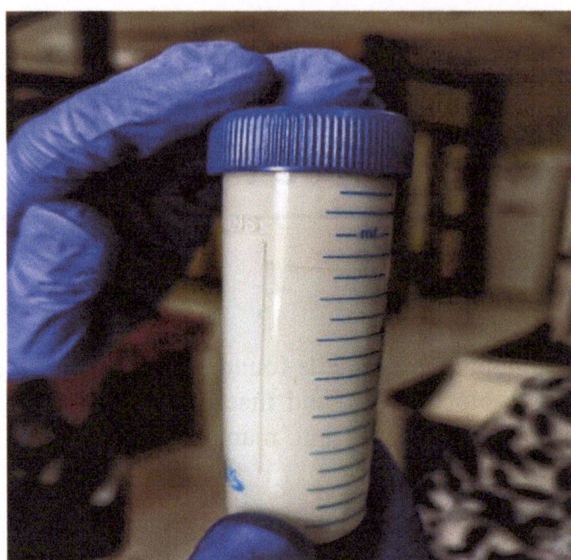

Fig. 12: Asceptical collection of milk in a 50 mL sterile tube.

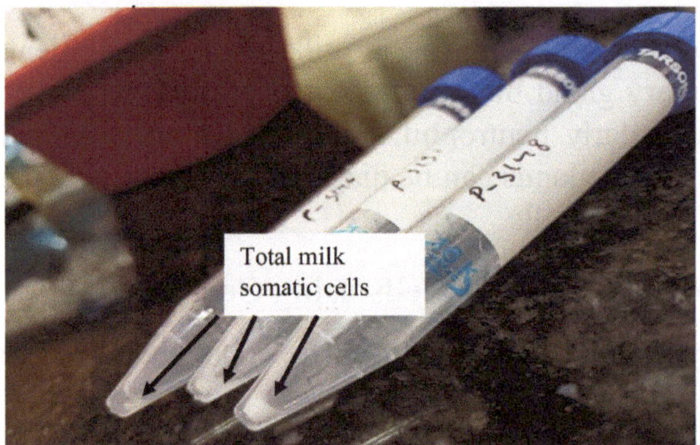

Fig. 1: Milk somatic cells isolated from buffalo milk samples. The figure shows varying cell pellet sizes, which reflect differences in somatic cell counts across the samples. These variations are influenced by individual buffalo differences, lactation stage, parity, and breed.

Further Readings

Caplan, Z., Melilli, C and Barbano, D. M. (2013). Gravity separation of fat, somatic cells, and bacteria in raw and pasteurized milks. J. Dairy Sci, 96(4), 2011-2019.

Widmer, J., Desclous, L., Brugger, C., Jager, M-L, Jager, Berger, T. and, Egger, L. (2022). Direct labeling of milk cells without centrifugation for counting total and differential somatic cells using flow cytometry. J. Dairy Sci, 105(11), 8705-8717.

10

Staining of Milk Somatic Cells

Background

Staining of milk somatic cells is the oldest method of enumerating cell types in milk by direct microscopic examination. Various stains, namely propidium iodide, methylene blue and hematoxylin-eosin stains, are used.

The total milk cell (TMC) pellet obtained above was resuspended gently in 500 µl of PBS 1X (Sigma, MA, USA). Place 20 µl of cell suspension on a clean glass slide (BioGenex, CA, USA) and let it dry at room temperature.

Materials and reagents required

1. Positively-charged glass slides
2. Glass coverslips
3. Hot plate
4. Fixatives – Methanol or ethanol
5. Hematoxyline and Eosin stain
6. Phosphate-buffered saline (PBS)
7. Distill water
8. Ethanol for dehydration
9. Xylene
10. Mounting medium (like DBX)

Procedures

1. Smears of milk somatic cells are first air-dried at room temperature
2. After air dry, fix smears on hot plate at 60 °C for 1 h.
3. After heat fixation, cells are again fixed with methanol for 10 min followed by air dry.
4. Hydration: Hydrate the fixed cell smear by giving 10 dips X 3 times in distilled water.
5. Color nuclei: Stain in hematoxylin for 6 minutes.

6. Wash the hematoxylin-stained smear by giving 10 dips X 3 times in distilled water.
7. Differentiation: Give 10 dips in acid alcohol (0.5% HCl in 70% ethanol) to remove an extra stain.
8. Blueing: Rinse slide in distilled water by giving 10 dips X 4 times.
9. Counterstain: Give 2 dips in Eosin (0.5% w/v).
10. Wash cell smear by giving 10 dips X 3 times in distilled water.
11. Dehydration: Dehydrate in increasing concentration of alcohol.
12. Clearing: Give 10 dips in xylene (Sigma, CA, USA) with three changes.
13. Mounting: Mount in DPX (Himedia, Maharashtra, India) and carefully place coverslips (Bio-Genex, CA, USA).
14. Observe under a microscope (Figure 14) and capture photomicrographs for the analysis.

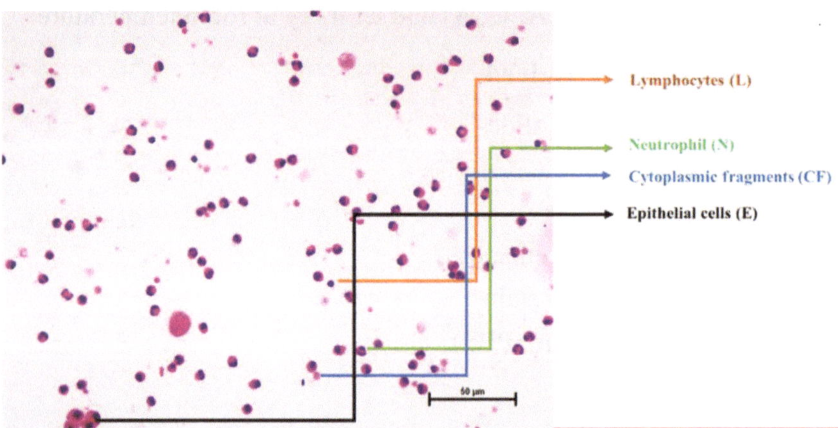

Fig. 14: Hematoxylin and Eosin (H&E) Staining of milk somatic cells. The image shows the differential staining of cellular structures, with nuclei stained blue/purple by hematoxylin and cytoplasm-stained pink by Eosin. These stained cells help identify cellular morphology and assess the presence of inflammatory cells, epithelial cells, and cytoplasmic fragments.

Note: Staining milk somatic cell counts is not a reliable method of counting milk cells. Enumeration of milk somatic counts is done by the Broadhurst Paley stain method and, more recently, using automated milk somatic cell counters like DeLaval cell counter DCC, Sweden (Jose et al., 2022; DeLaval. 2024).

Further Readings

DeLaval. 2024. Available: https://www.delaval.com/en-in/milking-essentials/milking-accesories/occ-dcc/delaval-cell-counter-dcc/ Accessed Oct 17, 2024.

Jose, KR, Vijaykumar, K. Justin Davis K and Shyma, VH (2022). Evaluation of two different methods to estimate the somatic cell count of bovine milk. J. Indian Vet. Assoc. 20 (1): 39-47.

11

Immunostaining of Milk Somatic Cells

Background

Immunostaining is a technique used to visualize specific proteins or antigens within cells. In milk somatic cells, immunostaining is employed to identify and quantify various cell types, study their functional characteristics, and diagnose diseases. Immunostaining can be used to identify and characterize different cell types based on the expression of specific surface markers. For example, CD45 is a leukocyte marker, while cytokeratins are markers for epithelial cells. In addition to the research on mammary gland biology, immunostaining is a valuable tool for studying the biology of the mammary gland. It can be used to investigate the expression of genes involved in milk production, cell differentiation, disease processes and even pregnancy detection. For example, galectin-3-binding protein (LGALS3BP) is a gene with elevated expression in the PBMCs of early pregnant buffaloes (Sharma et al., 2023). In this chapter, we have investigated the protein expression of LGALS3BP in milk somatic cells to determine if immunopositive cells are present in milk. Apart from this, stem cells in milk can be identified using markers like CD133, CD73, CD44, and other cell surface markers (Coni et al., 2023).

Materials and Reagents Required

1. Fixative- cold acetone, ethanol, or aldehyde-based fixative (10% neutral buffered formalin/ NBF or paraformaldehyde).
2. Ethanol – 70%, 90% and 100%
3. Xylene
4. Washing buffer (phosphate buffer saline)- PBS
5. Hydrogen peroxide will be required for endogenous peroxide blocking in the cells in the case of the horseradish peroxidase (HRP) enzyme-based color detection system..
6. Normal horse serum or any other serum: 2.5% horse serum or any other serum can be used to block non-specific epitopes.

7. Primary antibody of your choice- Select specific antibodies is required. Ask the vendor or see the literature to see if your primary antibody is reactive to the intended species.
8. Secondary antibody
9. HRP-IHC kit (like Super Sensitive™ Polymer HRP Kit Manual; Biogenex) - If you use the kit, follow the manufacturer's instructions (Figure 15).

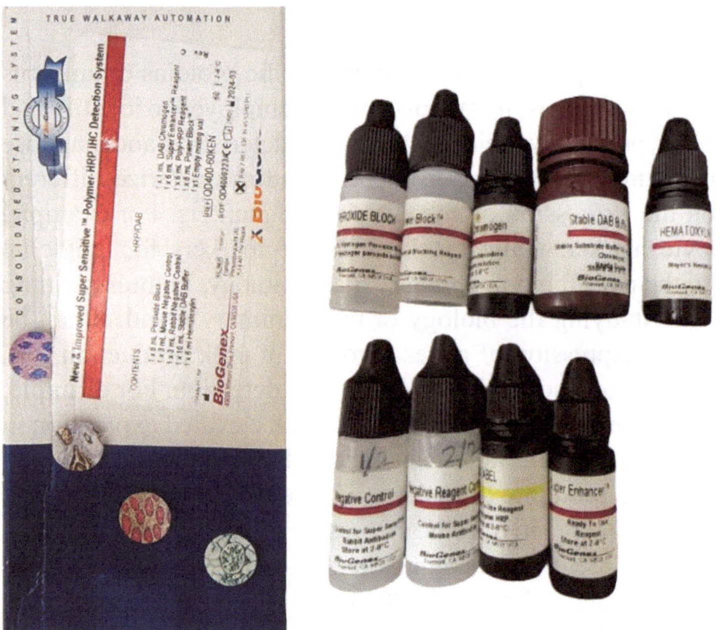

Fig. 15: Components of the Super Sensitive™ Polymer-HRP IHC Detection Kit. The kit contains various components and enables enhanced sensitivity and specificity in immunohistochemistry (IHC) assays.

Procedures

Rehydration

- Rehydrate the cell smear in 1X PBS (Sigma, CA, USA) for 5 minutes.

Endogenous Peroxidase Blocking

- 3% H2O2 is used for 20 minutes.

Protein Blocking

- Cover the cell smear with a 5% goat serum or casein block blocking agent.

- Cover it with a plastic coverslip (shown right). Incubate 1 h at RT for blocking.

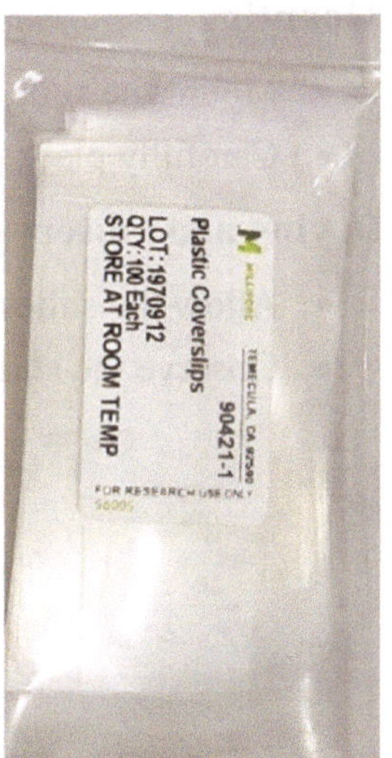

Primary Antibody Incubation

- Cover the cell smear with the primary antibody (for example, monoclonal antibody LGALS3BP at 1:500 dilution; BIO-RAD, CA, USA).
- Incubate for 1-2 h at room temperature.

Washing

- Wash the smear with PBS-T for 5 minutes; repeat 3 times.

Secondary Antibody Incubation

- Cover the cell smear with 3-4 drops of universal secondary antibody (anti- mouse/rabbit IgG; Vector, NY, USA).
- Incubate at room temperature for 30 minutes.

Washing

- Wash the smear with PBS-T for 5 minutes, repeat 3 times.

DAB Staining

- Add 100 µl of DAB (Immunotag, MO, USA) over the cell smear.
- Incubate in the dark for 3-5 minutes.

Stop Reaction

- Stop the reaction by rinsing the slide with distilled water.

Hematoxylin Counterstaining

- Immerse the slide in hematoxylin for 5 minutes.

Washing

- Wash the slide with distilled water using ten dips and repeat 3 times.

Dehydration

- Dehydrate the slide with three changes of 100% ethanol.

Clearing

- Clear the slide using three changes of xylene (Sigma, CA, USA).

Mounting

- Mount the slide using DPX (Himedia, Maharastra, India).
- Carefully place coverslips (Bio-Genex, CA, USA).

Drying and Observation

- Allow the slides to dry.
- Observe the slides under the microscope for the Staining (Figure 16).

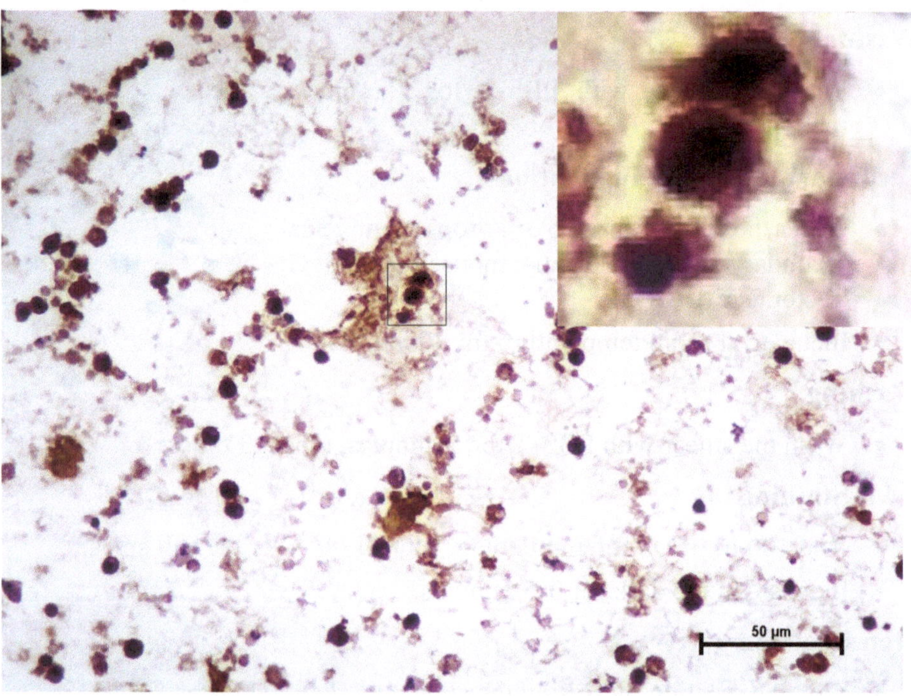

Fig. 16: LGALS3BP-immunostained milk somatic cells. The cells were stained with a monoclonal antibody against LGALS3BP (galectin-3 binding protein) followed by DAB chromogen visualization, resulting in brown staining of positive cells. Hematoxylin was used as a counterstain to highlight cell nuclei in blue. Inset- magnified image of immunopositive cells.

Further Readings

Coni P, Piras M, Piludu M, et al.)2023). Exploring cell surface markers and cell-cell interactions of human breast milk stem cells. Journal of Public Health Research. 2023;12(1)

Sharma P, Choudhary RK, Ratta NS, Singh ST. Investigation of conceptus stimulated gene expression in buffalo peripheral blood mononuclear cells as potential diagnostic markers of early pregnancy. Journal of Dairy Research. 2023;90(2):142-145.

Section D
Fine Needle Aspirate

12

Preparation of FNA Smear

Background

During a fine needle aspiration biopsy (FNAB), a small amount of mammary tissue fluid is removed using a fine needle by applying suction, and cells present therein are identified in a thin smear. FNAB has been the most sensitive, quick, and precise tool for the differential diagnosis of malignancy. However, further research is required into the possibility of malignancy in dogs based on cytological evidence and candidate gene expression analysis. This is a non-invasive cancer detection technique and is often used in clinics.

Materials and Reagents Required

1. Fixative- cold acetone, ethanol, or aldehyde-based fixative (10% neutral buffered formalin/ NBF or paraformaldehyde).
2. Ethanol – 70%, 90% and 100%
3. Washing buffer (phosphate buffer saline)- PBS
4. Clean glass slides – preferably positively charged
5. 22-gauge needle fitted with 2 ml syringe

Procedures

Prepare for a dog having a mammary tumor

- Clean the site of the mammary tumor with alcohol and allow it to dry.
- Keep slides ready, clean, and dry.

Obtain the FNA Sample

- Attach a syringe to the needle.
- Insert the needle into the tumor, applying gentle suction to collect cells. Alternatively, during mastectomy, the mass of the dissected tumor can be used to draw FNA (Figure 17).

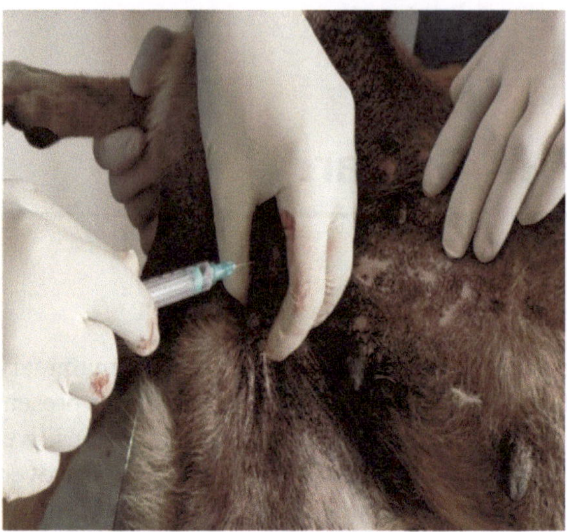

Fig. 17: Palpation of the mammary tumor to raise the mass and insertion of the needle with back-and-forth motion to collect sufficient FNA from tissue.

Prepare the Slide

- Place a clean, dry slide on a flat surface.
- Place 1-2 drops of collected FNA on the slide surface

Create the Smear

- **Method 1**
  - Place a second clean slide and sandwich sample drop and make a smear.
  - Gently pull the top slide backward, creating a thin, even smear. Usually, an oval-shaped smear is ideal.
- **Method 2:**
  - Simply keep 10- 50 µl of drop on the slide and let it dry as such.

Air Dry

- Allow the slide to air dry completely.

Fixation

- Immediately after air drying, immerse the slide in fixative for the recommended time to preserve the cellular details.

When slides are dry, they can be stored at room temperature for staining.

Further Readings

Luhach, P. (2022). Investigation of fine needle aspirates from canine mammary tumors – a translational model for diagnosing human breast cancer (MSc thesis). Guru Angad Dev Veterinary and Animal Sciences University, Ludhiana.

Yu, Y. H., Wei, W., & Liu, J. L. (2012). Diagnostic value of fine-needle aspiration biopsy for breast mass: a systematic review and meta-analysis. BMC cancer, 12(1), 1-14.

13

Staining of FNA Smears Giemsa Staining

Background

Staining is a crucial step in the microscopic examination of fine needle aspirates (FNAs) from canine mammary tumors. It enhances the visibility of cellular structures, making it easier to diagnose the tumor type. The cryopreservation of unfixed, air-dried smears of FNA can safely be used for staining up to 33 months, including immunostaining for epithelial and mesenchymal cell markers like cytokeratin and vimentin (Furusawa et al., 2023).

Common Staining Techniques

1. **Romanowsky Stains:** Romanowsky stains are a group of related stains, Dimitri Romanowsky in 1891. It used a mixture of Eosin (an acidic dye) and methylene blue (a basic dye). The original method has since been modified to include additional derivatives of methylene blue, such as azure B, leading to more distinct staining effects (Marshall, 1986). It is widely used in hematology, cytology, and microbiology for staining blood smears, bone marrow smears, and other cell types. These stains are significant for identifying and differentiating cellular components like nuclei, cytoplasm, and intracellular structures in various types of cells.
   - **May-Grunwald-Giemsa stain:** A modification of the Romanowsky stain, combined with Giemsa for its ability to differentiate between cell types and structures. It stains nuclei blue and cytoplasm pink. Giemsa stain utilizes the azure dye to stain the azurophilic cellular components purple.

Materials needed for May-Grunwald-Giemsa stain

1. May-Grünwald Stain (Cat no. 205435; Sigma) – Take 0.25 g of May-Grünwald powder in 100 mL of methanol to make 0.25% (W/V) May-Grünwald stock solution.

2. Giemsa Stain (Cat no. G5637; Sigma)- 3.8 g of Giemsa powder + 250 ml methanol + 250 ml glycerol. This would be Giemsa's stock solution. The final concentration of this Giemsa stock solution will be 0.76% w/v (3.8 g of Giemsa powder in 500 mL of solvent).
3. Buffer Solution (pH 6.8)- PBS
4. Distilled Water
5. Glass Slides
6. Microscope
7. Staining Jar or Coplin Jars

Procedures

- Dilute the stock May-Grünwald (0.25% w/v in methanol) stain 1:1 with phosphate buffer (pH 6.8). For example, mix 10 ml of stock solution with 10 ml of PBS to make 20 ml of working solution of May-Grünwald.
- Dilute the stock Giemsa solution (0.76% w/v in methanol and glycerol) stain 1:10 with phosphate buffer (pH 6.8). For example, mix 2 ml of Giemsa stock solution with 18 ml of PBS to make 20 ml of working solution of Giemsa.
- Air-dry FNA smear on glass slides and fix them in methanol for 3 -5 min.
- Stain smear for 3 to 5 min while working on the May-Grunwald solution.
- Add an equal volume of PBS directly to the slide without discarding the May- Grünwald solution to dilute it- wait for 5 min.
- Rinse in PBS - 2 x 3 min
- Stain for 15 - 20 min in working Giemsa solution.
- Rinse in PBS – 2 x 3 min.
- Air dry and mount as desired.

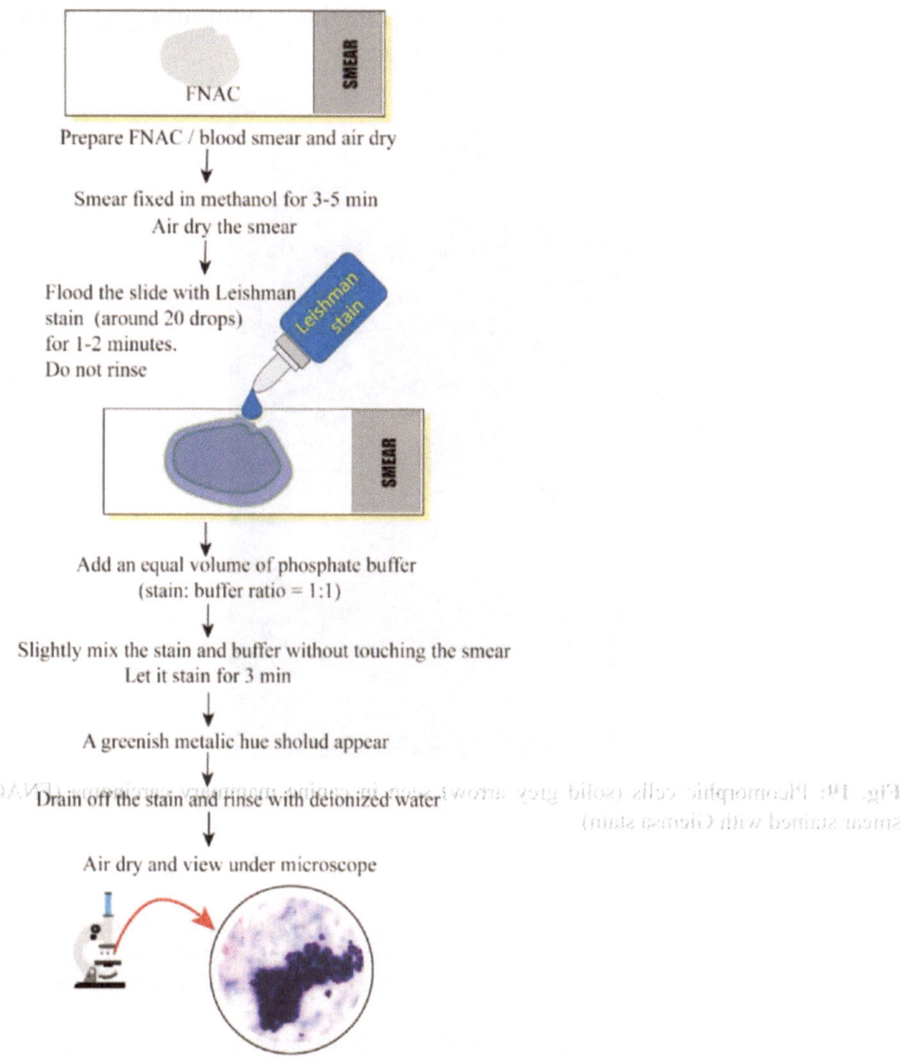

Fig. 18: Schematic diagram of Giemsa staining of FNA cytology smear.

Results and Interpretation

The following distinct features are seen in Giemsa-stained cells (Figure 19).

- Nuclei stain blue and purple (curved arrows), while the cytoplasm stains pink to orange.
- Erythrocytes: Stain pink (stars).
- Granulocytes: Show specific granules

- Abnormal cells (like cancer cells) might exhibit changed nuclear morphology or intensified dyeing intensity.

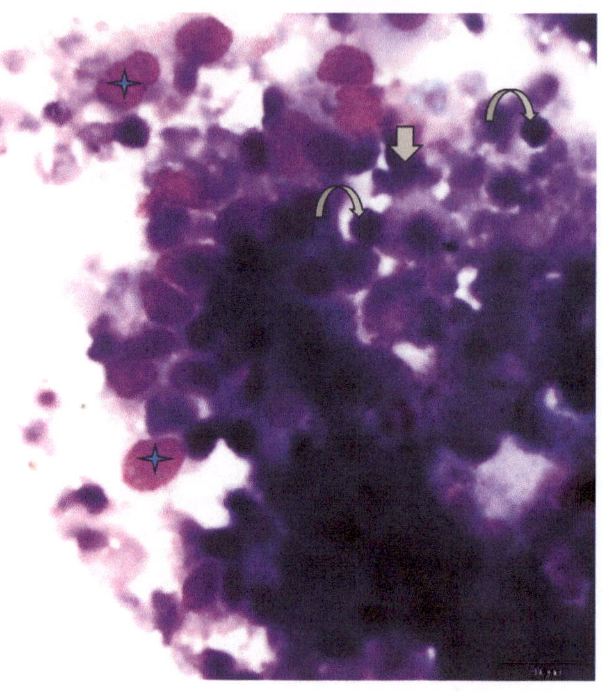

Fig. 19: Pleomorphic cells (solid grey arrow) seen in canine mammary carcinoma (FNAC smear stained with Giemsa stain)

Further Readings

Furusawa, Y., Shima-Sawa, M., Hifumi, T., Miyoshi, N., Yamato, O., & Yabuki, A. (2023). Usefulness and Limitations of Cryopreservation for Immunocytochemical Staining of Canine Cytological Specimens for Detection of Cytokeratin and Vimentin. Veterinary Sciences, 10(2), 153.

Lillie, R.D. (1977). H.J. Conn's Biological Stains: A Handbook on the Nature and Uses of the Dyes Employed in the Biological Laboratory. Williams & Wilkins.

Marshall, P.N. (1986). Staining of Blood and Bone Marrow Films. Wright & Sons.

14

Staining of FNA smears Leishman Staining

Background

Leishman stain (named after W. B. Leishman, a British surgeon) is one of the Romanowsky- type stains. It is a differential stain used to stain the different components of the cells variably. The Leishman stain is one of the best for routine blood staining and can be used for FNAC smear staining. It stains the peripheral blood smear when examining blood film under the microscope (Figure 20).

Leishman stain combines Azure B (blue) and Eosin Y (orange) dyes. Azure B has an affinity for DNA and acidic groups in cell nuclei, while Eosin Y has an affinity to basic groupings in cytoplasm. The staining results show crisp blue nuclei while the cytoplasm takes a purple hue.

Procedure for Leishman Staining

Sample Preparation

- Obtain a thin blood smear or FNAC sample.
- Air-dry the smear thoroughly.
- Flood the slide with methanol for 3-5 minutes for fixation
- Air dry the smear.

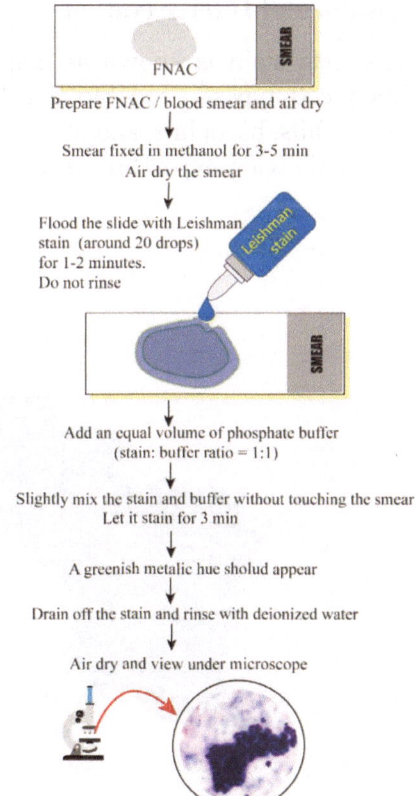

Fig. 20: Pictorial representation of protocol for Leishman staining.

Staining Steps

1. Flooding with Leishman Stain: Flood the slide with Leishman stain (around 20 drops) for 1-2 minutes. Do not rinse.
2. Adding Phosphate Buffer:
   i. Add an equal volume of phosphate buffer (stain: buffer ratio = 1:1). ii. Slightly mix the stain and buffer without touching the smear.
   iii. Let it stain for 3 minutes. A greenish metallic hue should appear.
3. Rinsing and Drying:
   i. Drain off the stain buffer solution.
   ii. Rinse the slide with deionized water.
   iii. Use a hair dryer on low speed to completely dry the slide. iv. Air dry the smear before the examination.

Results and Interpretation

Leishman stain is known as diagnostic gold. It helps differentiate various blood cell types— red blood cells, lymphocytes, monocytes, neutrophils, eosinophils, basophils, and platelets. Tumor cells present in FNAC are seen in cluster showing hyperchromatic and pleomorphic nuclei (Figure 21).

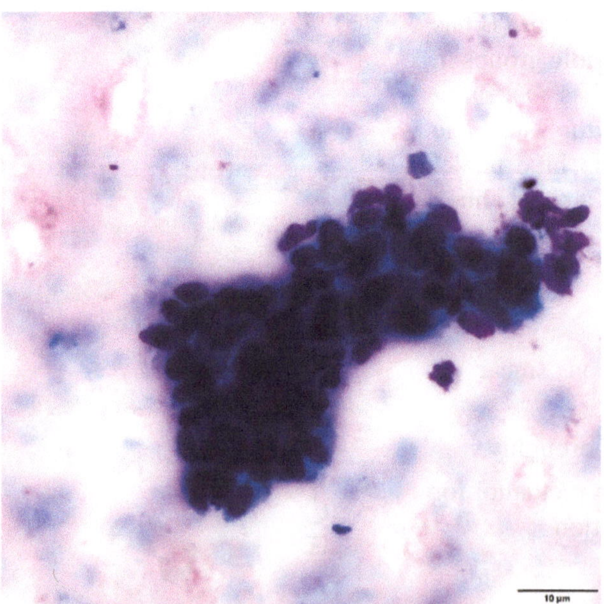

Fig. 21: Pleomorphic and hyperchromatic cells in clusters can be seen in canine mammary carcinoma. Cell smear was made using FNAC of confirmed case of canine mammary tumor.

15

Immunostaining of FNA Smears for the Identification of Proliferating Cells

Materials and Reagents Required

1. Fixative- cold acetone or methanol
2. Ethanol – 70%, 90% and 100% for rehydration and dehydration
3. Xylene for clearing
4. Washing buffer (phosphate buffer saline)- PBS
5. Hydrogen peroxide – for endogenous peroxidase blocking
6. Normal horse serum or any other serum**:** 2.5% horse serum or any other serum can be used to block non-specific epitopes.
7. Ki67 mouse primary antibody
8. Secondary antiboy kit - Impress Excel kit (Cat no. MP-7601; Vector Laboratories, NY, USA) or a similar kit from other vendors like Biogenex (Figure 22; left panel).
9. Peroxide substrate kit—Immpact NovaRed (cat no. SK-4850; Vector Laboratories, NY, USA) for developing a red color signal (Figure 22; right panel).

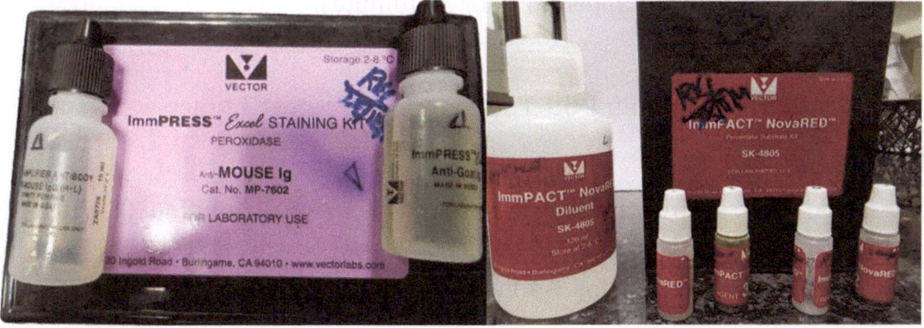

Fig. 22: Anti-mouse HRP-conjugated secondary antibody kit (left panel) and ImmPACT NovaRed peroxide substrate kit (right panel).

Procedures

Rehydration

- Rehydrate the cell smear in 1X PBS for 5 min.

Endogenous Peroxidase Blocking

- 3% H2O2 - 20 minutes (kit contains one vial of hydrogen peroxide in the name of ”Peroxoblock”)

Protein Blocking

- Cover the cell smear with 2.5 % horse serum (kit contains one vial of hydrogen peroxide in the name of “Bloxall”)
- Incubate for 30 min at RT for blocking

Primary Antibody Incubation

- Cover the cell smear with the primary antibody (cell proliferation marker anti- mouse Ki67 ready to use (RTU) (Biogenex, CA USA).
- Incubate for 1-2 h at room temperature.

Washing

- Wash the smear with PBS-T for 5 minutes, repeat 3 times.

Secondary Antibody Incubation

- Cover the cell smear with 3-4 drops of secondary antibody (anti-mouse Ig; Vector Lab, NY, USA).
- Incubate at room temperature for 30 minutes.

Washing

- Wash the smear with PBS-T for 5 minutes, repeat 3 times.

DAB Staining

- Add 100 µl of working ImmPACT NovaRED solution (1 drop of concentrate to 1 ml of ImmPACT NovaRED diluent) over the cell smear.
- Incubate in the dark for 3-5 minutes.

Stop Reaction

- Stop the reaction by rinsing the slide with distilled water.

Hematoxylin Counterstaining

- Immerse the slide in hematoxylin for 5 minutes.

Washing

- Wash the slide with distilled water using ten dips and repeat 3 times.

Dehydration

- Dehydrate the slide with three changes of ascending grades of ethanol.

Clearing

- Clear the slide in three changes of xylene

Mounting

- Mount the slide using DPX (Himedia, Maharastra, India).
- Carefully place coverslips (Biogenex, CA, USA).

Drying and Observation

- Allow the slides to dry.
- Observe the slides under the microscope for the Staining (Figure 23).

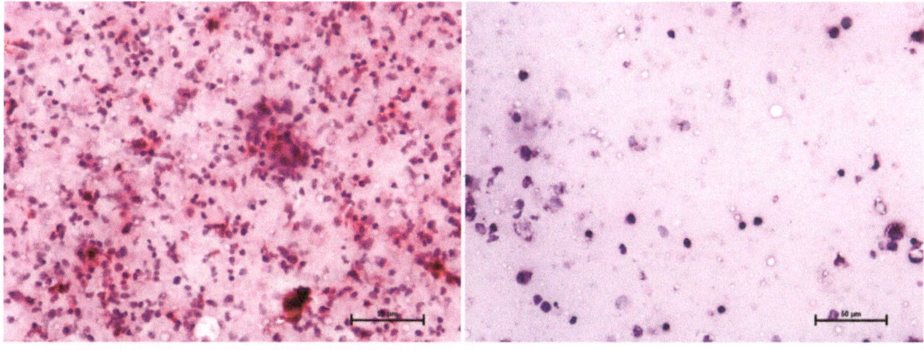

Fig.23: Ki-67 immunostaining of FNA smears from the canine mammary tumor (left panel) and healthy mammary gland (right panel). The FNA smear of the tumor (left) shows increased cellularity, with dense clusters of abnormal or malignant cells, often arranged in disorganized groups. In contrast, the healthy mammary gland smear (right) displays low cellularity, with few inflammatory cells. The tumor smear also exhibits increased Ki-67 expression, indicating a higher proliferation rate than healthy tissue.

Further Readings

Vector lab Immunohistochemistry work flow https://vectorlabs.com/applications/immunohistochemistry-workflow?srsltid=AfmBOopxffXTwXqVNagA05mRlixMWSyIg33dNtcSbfe1m0zu Ws497j14 (assessed on Aug 21 2024)

Soliman, N. A., & Yussif, S. M. (2016). Ki-67 as a prognostic marker according to breast cancer molecular subtype. Cancer Biology & Medicine, 13(4), 496–504.

Sahebali, S., Depuydt, C. E., Segers, K., Vereecken, A. J., Van Marck, E., & Bogers, J. J. (2003). Ki-67 immunocytochemistry in liquid based cervical cytology: useful as an adjunctive tool?. Journal of Clinical Pathology, 56(9), 681–686.

Section E
Analysis of Morphology of Cells

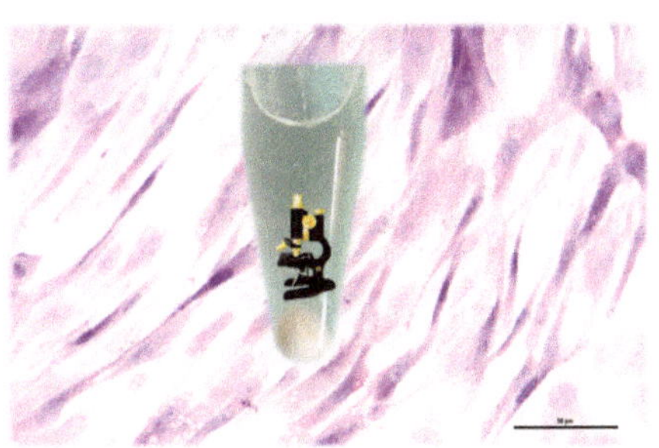

16

Introduction to ImageJ

Background

Cells are the basic unit of life. The morphology of cells describes their structural framework. It refers to the morphological features of the cell(s). It comprehends the various shapes and sizes in which cells function. Thus, the shape and size of the cell and the number of cells indicate their functional status. Hematological diagnostics requires the proper assessment and valid classification of cells for final diagnosis and prognosis. Clinicians, patients, and diagnosticians are all concerned with achieving a definitive specific diagnosis employing a single investigative test. The utilization of different cytological examinations from different organs gives adequate indicative data that leads to treatment outcomes.

What is ImageJ?

ImageJ is a public-domain image processing and analysis software that is Java-based and developed by the National Institutes of Health. The software is used to process and analyze scientific images.

The key features of ImageJ are 1) the versatility of working with various image formats, 2) extensibility with several plugins, 3) multidimensional data handling capability, and 4) vibrant community, making ImageJ a unique software for image processing and analysis. It can run as an online applet or be downloaded and used as an app.

How to Download and Install

Go to the ImageJ download page (magej.net/ij/download.html) and download for the operating system one has (Windows, Mac Os X or Linux) and install the application for your operating system (Figure 24).

home | docs | download | plugins | resources | list | links

Download

Platform Independent

To install ImageJ on a computer with Java pre-installed, or to upgrade to the latest full distribution (including macros, plugins and LUTs), download the **ZIP archive** (6MB) and extract the ImageJ directory. Use the *Help>Update ImageJ* command to upgrade to newer versions.

Mac OS X

Download **ImageJ bundled with Java 8** (may need to work around Path Randomization). Instructions. With M1 (ARM) Macs, download **ImageJ bundled with Zulu OpenJDK 13.0.6**.

Linux

Download **ImageJ bundled with Java 8**. Instructions.

Windows

Download **ImageJ bundled with 64-bit Java 8**. Instructions.

Fig. 24: Imagej can be download and installed for Mac, Linex or Window-based platforms

User interface

It has a menu bar containing all the menu commands, a toolbar, a status bar, and a progress bar (Figure 25). Images, histograms, profiles, widgets, etc., are displayed in additional windows. Most windows can be dragged around the screen and resized.

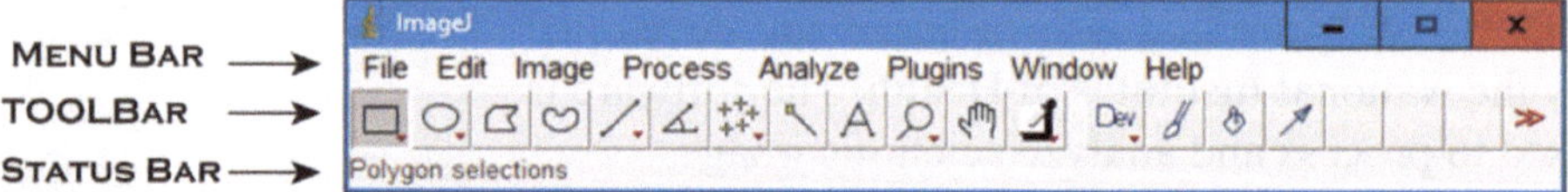

Fig. 2: Manu, tool and status bar of ImageJ after installation of software on the computer.

The Menu bar consists of file, edit, image, process, analyze, plugins, window, and help. The toolbar comprises rectangular, oval, polygonal, and freehand selection tools. After that, it comprises a straight line, angle tool, multipoint, wand tool, text tool, magnifying glass, scrolling tool, color picker, developer menu, paintbrush tool, flood fill tool and arrow tool.

Image processing

It offers various features like background correction intensity/density. It can calculate the area and pixel value statistics of user-defined selections. It can measure distances and angles. It can create density histograms and line profile plots. It supports standard image processing functions such as contrast manipulation, sharpening, smoothing, edge detection and median filtering.

a) Open Image in ImageJ

There are two ways an image can be opened in the ImageJ.

i. Through the File menu bar: Go to the file menu, then to open a submenu. Select the desired folder, then the desired image, and open it. The image will appear in a new window with the ImageJ heading (Figure 26).

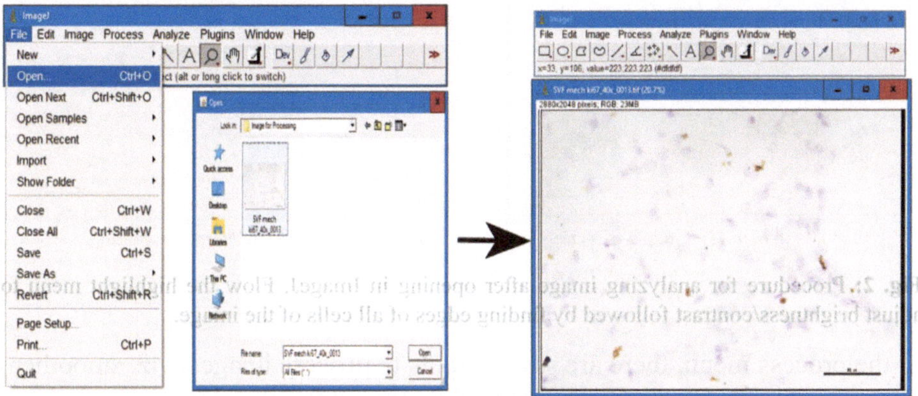

Fig. 2: Procedure for opening an image in Image for analysis through mouse click

File → Open→ Folder → image→ Open

ii. Through dragging the image: Open the folder and, select the desired image and drag it to the rectangle toolbar. The image will appear in a new window with the ImageJ heading.

b) Processing of the image

The image can be processed to enhance the quality of its image appearance. This is done using the image menu and the process menu. In the image menu, an image can be processed using the adjust submenu to change its brightness, contrast, window level, color balance, and many more functions. An example of image processing for finding edge of the cell (Figure 27) and improving quality using brightness/contrast (Figure 28) are shown.

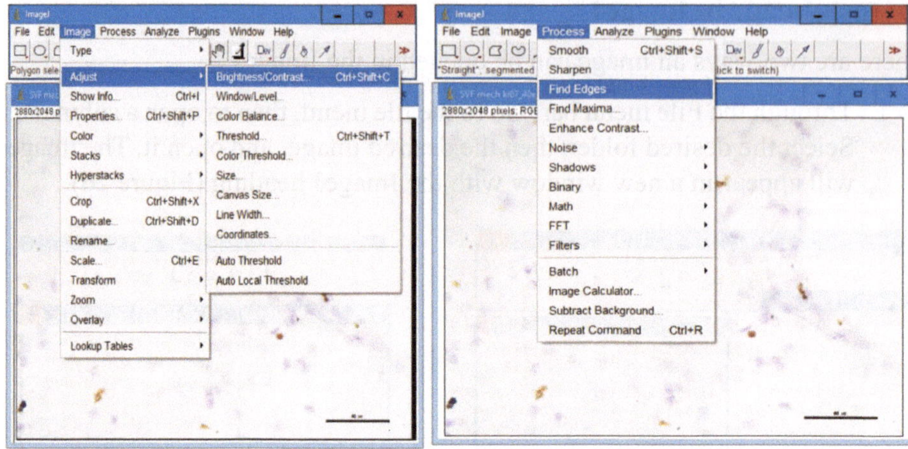

Fig. 2: **Procedure for analyzing image after opening in ImageJ. Flow the highlight menu to adjust brightness/contrast followed by finding edges of all cells of the image.**

In the process menu, there are several tools to process images, viz. smoothen, sharpen, enhance contrast, subtract background etc.

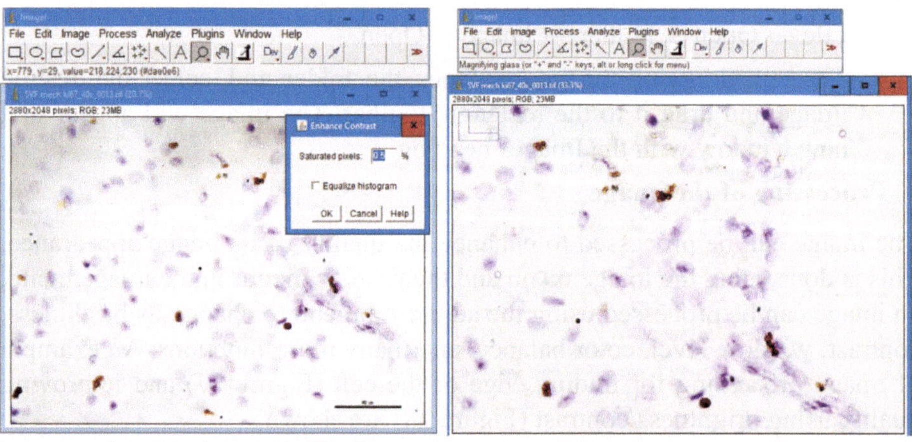

Fig. 28: **Procedure for enhancing brightness/contrast of the image showing before and after processing with the function. Improved quality of the image can be visualized after processing.**

Uses of ImageJ in Scientific image processing and analysis

ImageJ is a widely used tool in scientific image processing and analysis across various fields. Here are some key applications:

a) **General Image Processing:** ImageJ is used to enhance images, i.e., adjust brightness and contrast and apply filters to improve image quality. It is also used for segmentation, separating objects of interest from the background for further analysis.

b) **Microscopy Image Analysis:** It quantifies fluorescence intensity and colocalization studies and analyzes time-lapse sequences. It is used for 3D reconstruction and visualization of confocal stacks.

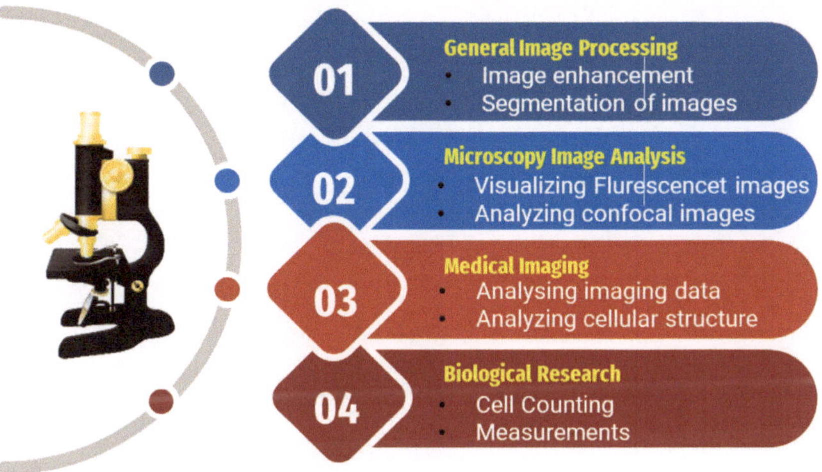

Fig. 2: Applications of ImageJ in scientific image processing and analysis.

c) **Medical Imaging:** It can also analyze and visualize medical imaging data, including segmentation and volume measurements. It is also used to quantify stained tissue sections and analyse cellular structures.

d) **Biological Research:** ImageJ is used to automate counting cells in culture or tissue samples and measure particles or cells' size, shape, and distribution.

Further Readings

Collins, T. J. (2007). ImageJ for microscopy. Biotechniques, 43(S1), S25-S30.

Hartig, S. M. (2013). Basic image analysis and manipulation in ImageJ. Current Protocols in Molecular Biology, 102(1), 14-15.

Sheffield, J. B. (2007). ImageJ, a useful tool for biological image processing and analysis. Microscopy and Microanalysis, 13(S02), 200-201.

17

Cell Counting using ImageJ

The cell number of a particular cell type in a total cell population might indicate the physiological or pathological status of the cell smear/FNAC/tissue. ImageJ is a powerful open- source software extensively used for image assessment, including cell counting. In this chapter, the steps to count cells using ImageJ has been discussed. There are two ways to count the cells in a research image using ImageJ.

A. Manual Counting

Use the "*Cell Counter*" plugin or simply use the multipoint tool in ImageJ. Open it, select the cell type you want to count, and click on each cell in the image to tally them. Double-click on the multipoint tool. A separate window will open named point tool. One can select the type of marking, color and size of making (Figure 30). Keep clicking on the cell type to be counted. It will record the number of cells counted. If one wants to count the second cell type in the same image, change the counter number and select the size, color and type of click point.

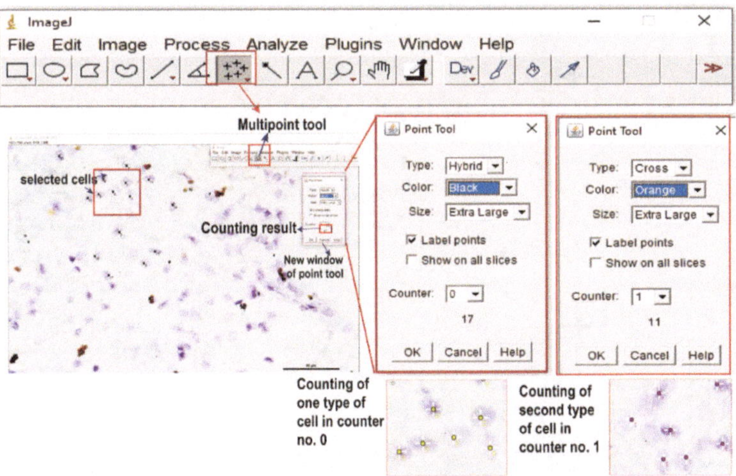

Fig. 30: Counting of cells using manual method (multipoint cell counter tool). The manual method is useful in processing of small number of images.

B. Automated counting

a) **Open Image in ImageJ:** First, open the image or image stack that contains the cells you want to count.

b) **Convert the image to 8-bit Grayscale (Optional):** Depending on your image type, one might need to convert it to an 8-bit grayscale image. One can do this by going to image> Type > 8-bit.

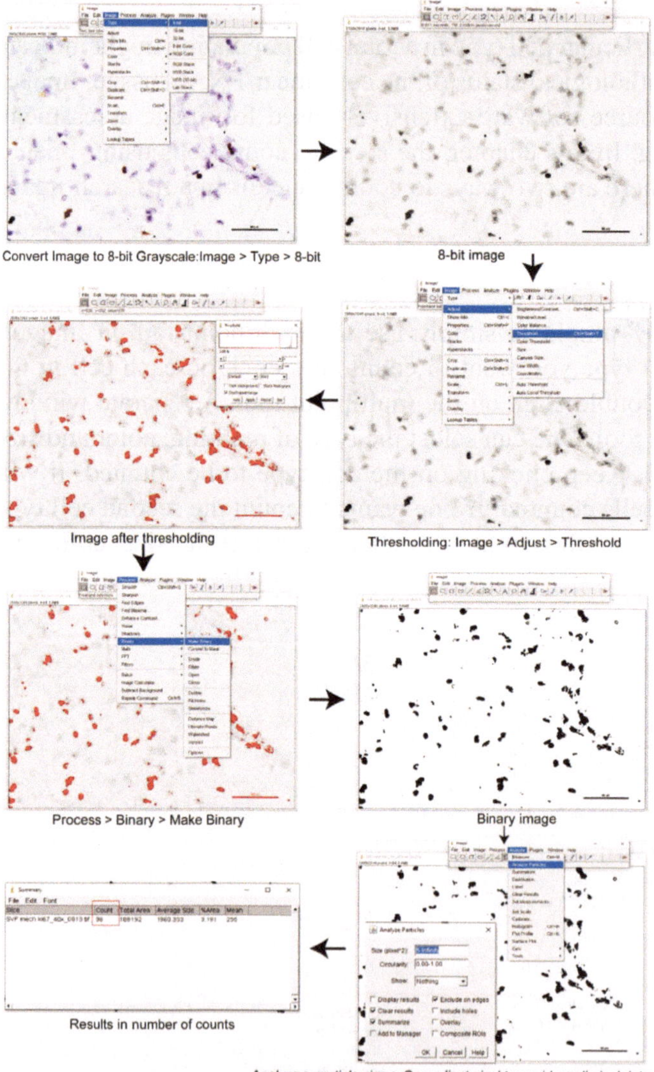

Fig. 31: Counting of cells using an automated method. The automated method is useful in processing of large number of images.

c) **Thresholding:** Thresholding helps distinguish cells from the background signal. Go to image> Adjust > Threshold. Adjust the threshold settings to ensure only the positively stained areas (cells) are visible. You can set the lower boundary to minimize noise.

d) **Convert to Binary:** After thresholding, convert the image to binary (black and white) using Process > Binary > Make Binary. This step simplifies cell identification.

e) **Separate Clusters (Optional):** If your cells are clustered, use Process > Binary > Watershed to separate them. Watershed segmentation helps split overlapping cells.

f) **Count the Cells:** If you have a large dataset, consider using automated methods. Plugins like Analyze Particles can automatically count cells based on size and intensity (Figure 31).

Note: Remember that the accuracy of cell counting depends on proper thresholding and segmentation. Adjust the parameters to get the best results for your specific images.

Further Readings

Grishagin, I. V. (2015). Automatic cell counting with ImageJ. Analytical Biochemistry, 473, 63-65.

O'Brien, J., Hayder, H., & Peng, C. (2016). Automated quantification and analysis of cell counting procedures using Image J Plugins. Journal of Visualized Experiments: JoVE, (117).

18

Cytomorphological Analysis of Cells

Measuring cell size, diameter, length, distance between cells, area, etc, are important indicators of the identification of cell types and the physiological or pathological state of cells and tissues. ImageJ is a useful program that allows researchers and scientists to work with microscopy images for cytomorphological detailing.

The steps to measure cell size are:

A. **Open Your Image in ImageJ:** First, open the image containing the cells you want to measure.

B. **Set the Scale (Optional):** If your image has a known scale (e.g., if you used a microscope with a calibrated scale bar), you can set it in ImageJ. This step ensures accurate measurements.

   Go to Analyze > Set Scale. Enter the known distance (in pixels) and the corresponding physical distance (e.g., micrometers). Click OK.

C. **Choose the Measurement Tool:** ImageJ provides several measurement tools. The most common ones for cell size are: Area Measurement: Use the Polygon Selection Tool to outline the cell, and then go to Analyze > Measure.

D. **Length Measurement:** Use the Straight-Line Selection Tool to measure the length of cell projections (e.g., neurites or filopodia).

E. **Record Measurements:** After selecting the appropriate tool, click and drag to measure the cell area or length. Press' t' to record the measurement. Repeat for multiple cells if needed.

F. **View Results:** The measured values will appear in the Results window.

   You can access the results by going to Analyze > Measure or by opening the Results window directly.

Note

Make sure to calibrate correctly to get accurate cell sizes. Remember that the accuracy of measurements depends on proper calibration and consistent use of the measurement tools. Consider using the area measurement method if you're dealing with irregularly shaped cells.

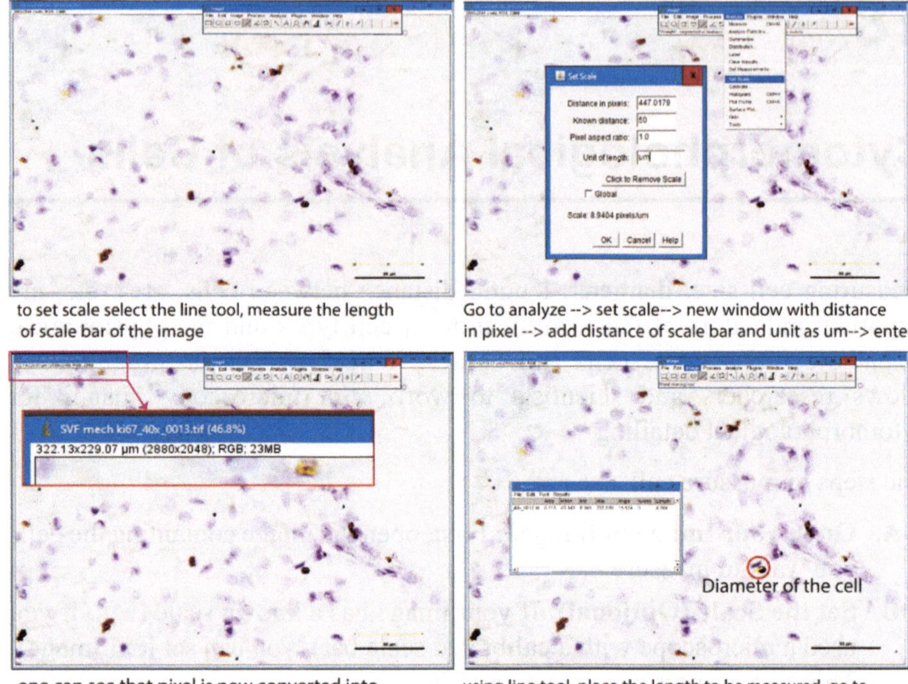

Fig. 32: Process of cytomorphological analysis of cells from the image. Example showing measuring the diameter of cells' nuclei. Cells are counterstained with hematoxylin and nuclei appearing blue. Proliferating cells, immunostained with Ki-67 showing brown-color nuclei.

Further Readings

Schindelin, J., Rueden, C. T., Hiner, M. C., & Eliceiri, K. W. (2015). The ImageJ ecosystem: An open platform for biomedical image analysis. Molecular Reproduction and Development, 82(7-8), 518-529.

Schneider, C. A., Rasband, W. S., & Eliceiri, K. W. (2012). NIH Image to ImageJ: 25 years of image analysis. Nature Methods, 9(7), 671-675.

Glossary

8-bit grayscale: When an image is scanned in black and white (greyscale), an 8-bit grayscale image captures 256 shades of gray (2 to the power of 8).

Acidic dyes: Negative, anionic, or acidic dyes like eosin, nigrosin, and Congo red have negative charges. They are repelled by bacterial cells, staining the background and leaving the bacteria bright and clear against a dark background for better vision and clarity.

Adhesion Proteins: Adhesion proteins, or cell adhesion molecules (CAMs), are located on the surface of cells and help cells interact with each other and their surroundings, like the extracellular matrix. These proteins are essential for maintaining the structure and proper function of tissues. In dogs, adhesion proteins are found in saliva, which may play a role in healing and tissue repair.

Antigen retrieval buffer: An antigen retrieval buffer is frequently used in IHC to expose antigens in formalin-fixed tissue sections. It can done by using Tris-EDTA or Citrate buffer in the presence of heat or enzymatic methods in the presence of proteinase K or DNase 1.

Antigen retrieval: Antigen retrieval is a technical step used on most formalin-fixed tissues during immunohistochemical staining. It helps restore the ability of antibodies to bind to epitopes, which can be lost during tissue fixation.

Antigen: An antigen is a foreign substance, like a protein or pollen, that triggers the body's immune system to produce antibodies.

Balanced salt solution (BSS): This buffer helps keep cells healthy by maintaining their natural pH level. It contains sodium chloride, potassium chloride, calcium chloride, sodium acetate, and sodium citrate.

Basic dyes: Basic dyes like methylene blue, toluidine blue, thionine, and crystal violet are positively charged compounds that stain negatively charged cell parts, like nucleic acids.

Biomarkers: Biomarkers are found in blood, other body fluids, or tissues. They are measurable indicators of a biological condition that provide information about any abnormalities or diseases.

Blast Cells: Blast cells are immature cells known as precursor or stem cells.

Blood Parasites: Organisms like Babesia and Ehrlichia are blood-infecting parasites that can sometimes be found in PBMC or whole blood smears.

BLOXALL: BLOXALL Blocking Solution is a dual enzyme blocker that inactivates peroxidase, pseudoperoxidase, and alkaline phosphatase in both formalin-fixed, paraffin-embedded, and frozen tissue sections.

Broadhurst Paley stain: The Broadhurst Paley stain method is used to measure the somatic cell count (SCC) in milk, which helps assess the mammary gland health of an Animal.

CD: Cluster of differentiation (CD) markers are cell surface proteins that are used to identify and classify cells, determining their development or maturity stage.

CD133: CD133 is a cell surface marker. Its expression is associated with stem cells and is used to identify and isolate cancer stem cells.

CD44: CD44 is a hyaluronan receptor found on all stromal cells and is a marker of stem cells. It also helps MSCs migrate through the extracellular matrix of normal tissues.

CD45: CD45 is a surface marker for hematopoietic cells and is used to identify and differentiate various types of cells

CD73: CD73 is a surface marker for identifying multipotent mesenchymal stem cells. It also tracks B cell differentiation, inflammatory bowel disease, and cancer.

Coccygeal vein: The coccygeal vein is found on a Bovine's tail and can be used for blood sampling and other procedures.

Collagen: Collagen is the primary structural protein in the extracellular matrix of the body's connective tissues.

Comparative Analysis: In vitro comparison of healthy and diseased PBMCs to examine biochemical pathways and evaluate the impact of external compounds.

Contamination: The presence of unwanted substances or bacteria in a sample, often occurring with non-sterile urine collection methods like free catch.

Cysteine: Cysteine is an amino acid present in our body with the chemical formula of

HO2CCH(NH2)CH2SH

Cystocentesis: An invasive technique where a needle is used to collect a sterile urine sample by going through the abdomen into the bladder. It helps test bacteria and cells.

Cytokeratin: Cytokeratins are a type of keratin protein present within the intracytoplasmic cytoskeleton of epithelial cells.

Cytokines: Cytokines are a broad category of small signaling proteins that act as chemical messengers. They help to control the immune response by performing specific functions.

Cytology: The examination of cells from urine under a microscope to diagnose diseases or infections.

Cytomorphological analysis: Cytomorphological analysis is the study of the morphology of cells.

Cytoplasmic localization: In IHC Cytoplasmic localization refers to the presence of target proteins in the cytoplasm, only the cytoplasm will be stained, and other areas will remain unstained.

DAB: 3,3'-Diaminobenzidine (DAB) is a common chromogen used in Immunohistochemistry and immunoblotting. In the presence of the enzyme peroxidase, DAB forms a brown precipitate that does not dissolve in alcohol or xylene.

Dehydration: In Immunohistochemistry (IHC), dehydration is the removal of water from tissue fixed in an aqueous solution.

DeLaval Cell Counter (DCC): DeLaval Cell Counter (DCC) is a portable device that measures somatic cell count (SCC) in cow milk by taking digital pictures of the cell nuclei and counting the cells automatically

Density gradient centrifugation: Density gradient centrifugation is a method used to separate different parts of whole blood based on their density, like PBMCs, granulocytes, platelets, and red blood cells. It uses a special solution, like Ficoll, to create layers that help isolate specific blood components.

Differential somatic cell count: Differential somatic cell count (DSCC) is a technique used to measure the percentage of polymorphonuclear leukocytes (PMN) and lymphocytes in milk.

DPX: Dibutylphthalate Polystyrene Xylene (DPX) is a clear, synthetic resin, nonaqueous mounting media used in IHC to fix sections on microscope slides. It dries quickly and helps preserve stains for a longer duration.

Enumeration: Enumeration is counting bacteria or microorganisms in any given biological sample.

Eosin: Eosin is a pink fluorescent xanthene dye that binds to positively charged molecules in the tissue. It is used in labs to stain proteins in cells and tissues. It is part of the hematoxylin and

eosin (H&E) staining method.

Epithelial cells: Epithelial cells line the inside and outside of the body, creating a protective barrier against infections. They are found in the skin, blood vessels, organs, and urinary tract areas.

Ethanol: Fixative ethanol is a chemical used in Immunohistochemistry that preserves tissues and proteins by dehydrating them and making proteins clump together

Ethylenediaminetetraacetic acid (EDTA): EDTA prevents blood samples from clotting. It is a chelating agent.

Fine-needle aspiration (FNA): Fine-needle aspiration is a diagnostic technique used to examine lumps or masses on the body. This minimally invasive biopsy procedure involves using a thin needle to collect a cell sample from a suspicious area in the body for further analysis.

Flow cytometry: Flow cytometry is an advanced method used to analyze and measure cells or particles' physical and chemical properties.

Fluorescence: Fluorescence is the absorbance of invisible radiation from certain substances, like ultraviolet light, and then emitting visible light.

Free Catch (Voided urine): A method of collecting urine as the animal naturally urinates.

Galectin-3 binding protein: Galectin-3 binding protein (galectin-3BP), encoded by the LGALS3BP gene, is a highly glycosylated protein that acts as a ligand for several galectins, including galectin-3. LGALS3BP has high expression in early pregnant buffaloes and may play a key role in early pregnancy recognition and implantation.

Genetic Testing (DNA Fingerprinting/Parentage Determination): Process for analyzing

DNA, either to identify individuals or determine parentage.

Giemsa stain: Giemsa stain is commonly used in histology, hematology, cytology, and bacteriology to stain blood cells, parasites, and other specimens It is composed of methylene blue, azure, and eosin.

Hematologic Abnormalities: Hematological disorders, also known as blood disorders, affect the shape or number of blood cells. These include conditions like anemia, bleeding disorders (e.g., hemophilia), blood clots, and blood cancers such as leukemia, lymphoma, leukopenia (low white blood cell count) or leukocytosis (high white blood cell count).

Hematologic Malignancies: Cancers related to blood cells, such as leukemia or lymphoma, that can be diagnosed using PBMC smears.

Hematology: Hematology is the study of blood and blood disorders

Hematoxylin: Hematoxylin is a dye used in pathology to color cells and tissues, especially in the H&E staining method. It comes from the heartwood of the logwood tree (Haematoxylum campechianum). As a basic dye, hematoxylin binds to negatively charged parts of the cell, like DNA and RNA, and gives the cell nucleus a blue or basophilic color.

Heparin: Heparin, also called unfractionated heparin, is a medication and naturally occurring glycosaminoglycan. It is a blood anticoagulant that increases the activity of antithrombin.

HiSep LSM 1077: HiSep LSM 1077 is a density gradient medium used in laboratory settings to isolate and purify blood cells, particularly peripheral blood mononuclear cells (PBMCs).

Histogram: A histogram is a visual representation of the distribution of quantitative data in graphical form.

Homeostasis: The process by which the body maintains internal stability, including the balance of substances in the mouth through saliva.

Hormone Assay: A test that measures hormone levels in the blood, often used in diagnostics and reproductive studies.

Horse Serum: is a ready-to-use solution for blocking non-specific binding or diluting antibodies in IHC. It is extracted from the blood of healthy horses. Normally. 2.5% Normal Horse Serum is used for protein blocking and as dilutent for primary antibody.

Horseradish peroxidase: Horseradish peroxidase (HRP) is a fast and stable enzyme used in various tests like IHC, western blot, and ELISA. It helps enhance strong signals quickly when attached to secondary antibodies.

Hyperchromatic nuclei: Hyperchromatic nuclei are dark-staining nuclei, indicating the presence of excess of DNA or chromatin.

ImageJ: ImageJ is a freely available Java-based image processing software program developed by the National Institutes of Health and the Laboratory for Optical and Computational Instrumentation. It is widely used in life sciences research for image processing, analysis, and editing of images.

ImmPACT NovaRED: The ImmPACT NovaRED HRP substrate creates a red color reaction and is 2-4 times more sensitive than the original SK-4800 substrate kit.

Immune-Mediated Disorders: **Immune-mediated diseases** occur when the immune system mistakenly attacks tissues or cells body's cells, which can be diagnosed through PBMC smear analysis.

Immunoglobulins: Immunoglobulins (Ig), are commonly referred to as antibodies, are specialized glycoproteins generated by plasma cells as part of the immune response. They play a crucial role in identifying and neutralizing foreign substances such as bacteria and viruses.

Immunomagnetic separation: Immunomagnetic separation (IMS) is a lab technique that uses magnetic beads coated with antibodies to quickly separate and concentrate specific cells or particles from a sample. The beads bind to the target, allowing it to be isolated using a magnetic field.

Immunopositive cells: Immunopositive cells are cells that react to a specific antigen or antibody producing specific positive results.

Immunostaining: Immunostaining is a laboratory technique that uses specific antibodies to identify particular proteins in a tissue or cell sample.

Inflammatory cells: Inflammatory cells are found in blood and tissue cells that play a major role in the body's immune response to injury, illness, or harmful substances like germs and toxins.

Jugular vein: The jugular vein is a blood vessel commonly used for intravenous (IV) drug administration, blood sampling, and other procedures.

Ki-67: Antigen Kiel 67, or Ki-67 or MKI67 (a marker of proliferation Kiel 67), **is a protein encoded by the MKI67 gene in humans** and provides information about how many cells are dividing.

Leukocytosis: An elevated white blood cell count, signaling infection or other conditions, observed in PBMC smears.

Lymphocytes: A Lymphocyte is a type of immune cell including T cells, B cells, and natural killer (NK) cells, essential for immune response. It is made in the bone marrow and is found in the blood and lymph tissue having two main types i.e., B lymphocytes and T lymphocytes. B lymphocytes make antibodies, and T lymphocytes help kill tumor cells.

Mastectomy: Mastectomy is a surgical process of removing all or part of a breast, including the skin and nipple.

Mastitis: Mastitis in Bovines is the inflammation of the udder caused by physical injury or infections from microorganisms. It is the most common and potentially fatal infection in dairy cattle worldwide.

Mastitis: Mastitis is a condition that leads to inflammation of the mammary gland and udder tissue in dairy cattle and is the main cause of reduced milk yield.

May-Grünwald-Giemsa (MGG) stain: The May-Grünwald-Giemsa (MGG) stain is the combination of May-Grünwald and Giemsa Romanowsky stain and is used in hematology and histology for highlighting blood and bone marrow cells.

Metabolic Toxicity Studies: Metabolic toxicity studies are a branch of life sciences that uses metabolomics to evaluate the risks and harmful effects of toxic substances on organisms.

Methanol: Methanol is a colorless liquid used to dehydrate and preserve tissues by fixing proteins. As the simplest alcohol, it is commonly used as a fixative for blood smears stained with Romanowsky stains. Methanol fixation works by denaturing and precipitating proteins, and as such it is a quick method

Methylene blue: Methylene blue also known as methylthioninium chloride is used in microbiology to stain animal, bacteria, and blood samples, highlighting dead cells and tissue structures for easier examination.

Microbiota: The community of microorganisms, including bacteria, present in urine, which can vary based on the method of collection.

Micturition Induction: The act of stimulating an animal, particularly rodents, to urinate by applying pressure to the abdomen or using physical intervention.

Micturition: The process of urinating, which can be induced in small animals by gentle pressure on the bladder.

Milk Epithelial cells: The epithelial cells in milk are mammary epithelial cells (MECs), which are responsible for milk production in the mammary glands of dairy cows and other mammals.

Milk Somatic cells: Milk Somatic cells (SCs) are a mix of immune cells and milk-producing cells that are released during milking. The concentration of SCs, referred to as the somatic cell count (SCC), serves as a crucial indicator of both milk quality and the health of dairy cows.

Moist chamber: A moist chamber is a container used to maintain moisture levels. It is commonly used during experiments.

Monocytes: A type of white blood cell in PBMCs that can differentiate into macrophages and dendritic cells, playing a role in immune defense and inflammation.

Monocytes: A type of white blood cell that differentiates into macrophages and dendritic cells, playing a role in immune defense and inflammation.

Multidimensional data: Multidimensional data is a dataset with many columns, known as features or attributes. Having more columns increases the chances of finding hidden insights.

Neoplastic Cells: Neoplastic cells are abnormal cells that grow and divide uncontrollably, forming a mass of tissue called a neoplasm or tumor.

Neutral buffered formalin: Neutral buffered formalin (NBF) is a formalin solution that is widely used to fix tissue samples in histology containing buffering agents to keep a stable pH.

Non-Invasive: A less stressful and painless procedure because it involves no cut or surgery.

Occupational disorders: Occupational disorder refers to an event or exposure to some toxic substances that occur in the workplace that cause or contribute to a condition or worsen a preexisting condition. They can be detected by testing PBMCs for immune or genetic changes.

Oral proteome: The complete set of proteins found in saliva, varying between species and linked to different physiological and pathological functions.

Palpation: Palpation is the examination process of diagnosing illness or disease using fingers or hands.

Parentage determination: Parentage determination uses DNA from blood (PBMCs) to establish genetic relationships, including parentage and evolutionary links.

Parity: Parity in animals refers to the number of times a female has given birth (Including stillbirths).

Parvovirus: A highly contagious virus in dogs that can be detected using saliva samples.

PBMC isolation: The method of separating PBMCs from other blood components is essential for studying them in experiments without interference from other blood cells.

Peripheral Blood Mononuclear Cells (PBMCs): PBMCs are a group of immune cells,

including lymphocytes (T cells, B cells, and NK cells), monocytes, and dendritic cells with round nuclei. They are isolated from whole blood for studying immune responses, disease mechanisms, and therapy development.

Peroxidase block: Peroxidase block is a step of Immunohistochemistry where a peroxidize reagent like Hydrogen peroxide (H_2O_2) is used to stop the natural peroxidase activity in certain tissues, like the liver, kidney, and red blood cells. This helps prevent unwanted background staining and false positive results.

Personalized medicine: It uses an individual's PBMCs to develop genetic immune profiles, enabling more effective and tailored treatments for individuals.

Pharmacokinetic (PK) study: PK analysis is done to determine the theory of how a drug behaves when it is administered *in vivo*.

Pleomorphic cells: Pleomorphic cells are cells in cytopathology and histology that vary in size, shape, or staining, or the size and shape of their nuclei.

Pleomorphic nuclei: Pleomorphic nuclei refer to variations in the size and shape of nuclei. **Pleomorphic**: In pathology, "pleomorphic" refers to a group of cells that vary widely in size, shape, or color.

Positively charged slides: These slides have a positive charge and are commonly used in Immunohistochemistry, which helps the negatively charged tissue stick firmly, preventing the tissue from lifting during staining.

Potassium permanganate: Potassium permanganate, with the chemical formula $KMnO_4$, is an oxidizing agent and is commonly used as a disinfectant, deodorizing, and astringent properties. It is also commonly referred to as Condy's crystals.

Propidium iodide: Propidium iodide (PI) is a commonly used fluorescent dye utilized particularly in flow cytometry to stain cells and nucleic acids (DNA and RNA).

Protein block: In IHC, protein blocking is done using a solution to prevent antibodies from attaching to non-specific sites of a tissue sample before applying the primary antibody. Serum is often used to block non-specific binding because it contains antibodies that bind to those sites. For best results, use a serum from the same species as the secondary antibody to avoid detecting non-specific binding.

Pyrogen-free: Pyrogen-free means the absence of pyrogens, substances that can cause fever. These pyrogens can come from bacterial endotoxins, or exotoxins, produced by the body and cannot even be killed during sterilization. AS sterilization kills living microorganisms but their endotoxins remain unaffected.

Rabies Virus: A virus that affects the nervous system and can be detected in dog saliva for diagnostic purposes.

RBC lysis: Red blood cell (RBC) lysis, also known as hemolysis, is the process of breaking open red blood cells and releasing their contents into the bloodstream

Regeneration and healing: Regeneration and healing are complex processes that allow organs and tissues to regain their original integrity and functionality after injury. It is supported by proteins like FGF (Fibroblast Growth Factor) and EGF (Epidermal Growth Factor), which are found in dog saliva and are related to tissue repair.

Rehydration: Rehydration is the process of removing wax from a tissue sample after deparaffinization to prepare it for immunohistochemistry staining.

Romanowsky stains: Romanowsky stains such as Wright, Giemsa, and Leishman are used in microscopy to identify cell types in blood and bone marrow. They mix acidic.

Saliva: A fluid secreted by salivary glands, which contains some inorganic substances, cells, microorganisms, etc. Saliva helps maintain oral health and aids in digestion.

Skim milk: Skim milk usually contains about 0.1% fat and is made by removing all the fat from whole milk.

Smear: A laboratory technique where a thin layer of PBMCs is spread on a slide for microscopic examination, aiding in disease detection.

Sterile Sample: A urine sample free from any contamination

Threshold limit: In ImageJ, a threshold limit is a cutoff value that separates an image into different pixel groups, like "foreground" and "background."

Toxicity Assessment: The process of testing how novel drugs interact with PBMCs to determine their safety and proper dosage for consumption.

Vacutainer: A Vacutainer is a sterile glass or plastic test tube with a colored rubber stopper creating a vacuum seal inside of the tube, used for blood collection, facilitating the drawing of a predetermined volume of liquid

Vimentin: Vimentin is used widely as a marker of the epithelial-to-mesenchymal transitions

(EMTs) that take place during embryogenesis and metastasis

Whole Blood: Blood collected directly from an organism, containing all components such as red blood cells, white blood cells, platelets, and plasma.